SKINCARE DECODED

SKINCARE DECODED

The Practical Guide to Beautiful Skin

REVISED & EXPANDED

VICTORIA FU & GLORIA LU

Skincare Chemists & Creators of Chemist Confessions

weldon**owen**

CONTENTS

SECTION 3: BUILDING YOUR ROUTINE

To the lost souls trying to figure out your skincare routine—
Read this book however you like, front to back,
back to front, or a section at a time. We hope it helps
you make some sense of the skincare world.
Enjoy!

Chemist Confessions is the accidental creation of two frustrated cubicle mates

The two of us met at L'Oreal, where we worked as skincare-formulation chemists. We quickly became friends because we share the same (terrible) sense of humor and the same frustrations with the industry. Thinking back, our whole crazy startup journey probably began as a little vent session in one of our cubicles.

We both felt like the beauty industry is just inundated with useless crap that gives consumers no way of navigating an overcrowded and overmarketed product landscape. Moreover, the chemist's voice (ya know . . . the people who actually put the formulas together?) was completely absent. Ultimately, we ended up leaving our comfortable corporate jobs, convinced that we'd probably leave the beauty industry entirely and move on to bigger and better things. During our time off, we created the Chemist Confessions Instagram account in late 2017 as a last hurrah to the industry. We wanted to share our insider knowledge in hopes that it would help someone find their next skincare product a little more easily.

When the account, against all odds, started gaining traction, we made a quick and perhaps foolhardy decision to jump headfirst into the startup world. Nowadays, we juggle the brand, the blog, the edu-Insta, and the podcast—all fueled by the belief that skin science should be fun, and that we all can obtain a beneficial skincare routine.

All in all, people say startup life is full of surprises. But this book was not even within the realm of what we could fathom creating. We used to joke that we pitied the poor future intern who would eventually have to catalog and organize our incredibly whimsical collection of Insta and blog posts. Joke's on us! This book was born from many nights of rereading all our old posts, researching, cringing at our evolving writing style, rewriting, drawing new sketches, brainstorming better ways to present difficult skin-science concepts, and just generally questioning our life choices. Oh, and alcohol. Lots of alcohol. Sorry, parents!

So, whether you've been a part of the Chemist Confessions journey all along, or are just discovering us through this book, we hope this book can help guide you in your next skincare purchase, with a few laughs along the way. And, as always, if you have any questions, write to us!

Your friendly neighborhood chemists,

Gloria & Victoria

Get in touch with us at
thebook@chemistconfessions.com

Section 1: BASICS

BASICS 101

There's no denying it—the skincare industry is a hot, hot mess! Toners, serums, essences . . . ampoules, creams, gel creams . . . lotions, masks—what does it all really mean?! And how do I even begin shopping for what I really need?! We feel you, so let's dissect these questions piece by piece. But before diving into products, let's cover all the fundamentals. With a good understanding of how your skin works, the different skin types, and an intro to the skincare-products landscape, you'll at least feel more comfortable with these overwhelming options. Welcome to Skincare 101!

SKIN BIOLOGY 101

Why, yes! A skincare book *does* always start with a diagram of the layers of your skin. Here's another one, except we're only highlighting the components that you'll encounter in important skincare-science concepts and common marketing lingo. Feel free to refer to this when you can't figure out what exactly that newly launched product is referring to.

Epidermis This is the outermost layer of skin and consists of cells called keratinocytes, which have a kind of weird, oval shape. Keratinocytes constantly regenerate and move "up" toward the outer layers.

Dermis The second major layer of skin. This is the structural layer holding everything "up."

Your skin is like an onion—it's got layers.

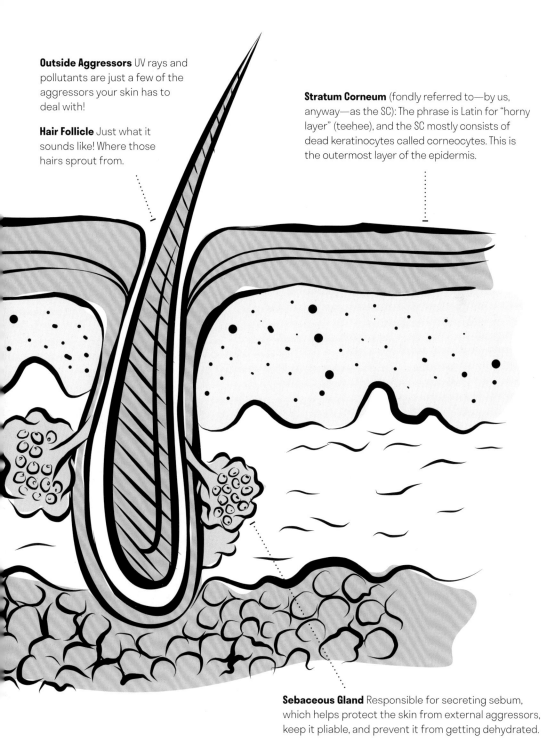

Outside Aggressors UV rays and pollutants are just a few of the aggressors your skin has to deal with!

Hair Follicle Just what it sounds like! Where those hairs sprout from.

Stratum Corneum (fondly referred to—by us, anyway—as the SC): The phrase is Latin for "horny layer" (teehee), and the SC mostly consists of dead keratinocytes called corneocytes. This is the outermost layer of the epidermis.

Sebaceous Gland Responsible for secreting sebum, which helps protect the skin from external aggressors, keep it pliable, and prevent it from getting dehydrated.

A CLOSER LOOK: EPIDERMIS

Now that we've established some basics, let's zoom in on the epidermis, as that's where most skincare products go to work.

Lipid Matrix A network of interlocking fatty substances comprised of ceramides, fatty acids, and cholesterol. The integrity of this hydrophobic (water-repelling) surface is crucial for sealing in moisture. Lipids in the matrix can be compromised due to environmental factors, skin diseases, and age. "Skin barrier care" products often refer to products aiming to protect lipid matrix health.

Hyaluronic Acid This popular skincare ingredient is actually naturally found in your skin in both the epidermis and dermis as a structural support and water grabber.

Acid Mantle & Microbiome Your SC has a slightly acidic pH that acts as a great defense system. Fun fact! The pH of your skin affects the little microbes living on its surface. There are theories out there that tie elevated skin pH to irritation and even acne.

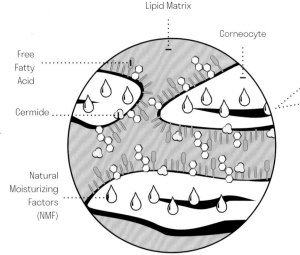

Lipid Matrix

Corneocyte

Free Fatty Acid

Cermide

Natural Moisturizing Factors (NMF)

Natural Moisturizing Factors (NMFs) Although your SC is mostly fatty, it requires some level of water to function correctly. Meet your NMFs. Molecules like urea, lactic acid, and sodium PCA act as your skin's naturally occurring water grabbers.

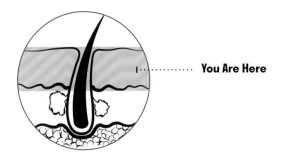

You Are Here

Melanocytes Responsible for the pigment in your skin. They produce melanosomes (small packets of pigment) as a protective, defensive reaction to sun damage. Sometimes they can get carried away, due to excessive sun damage and stress, and cause freckles and hyperpigmentation (dark spots). Your anti-dark-spots serum targets these little guys.

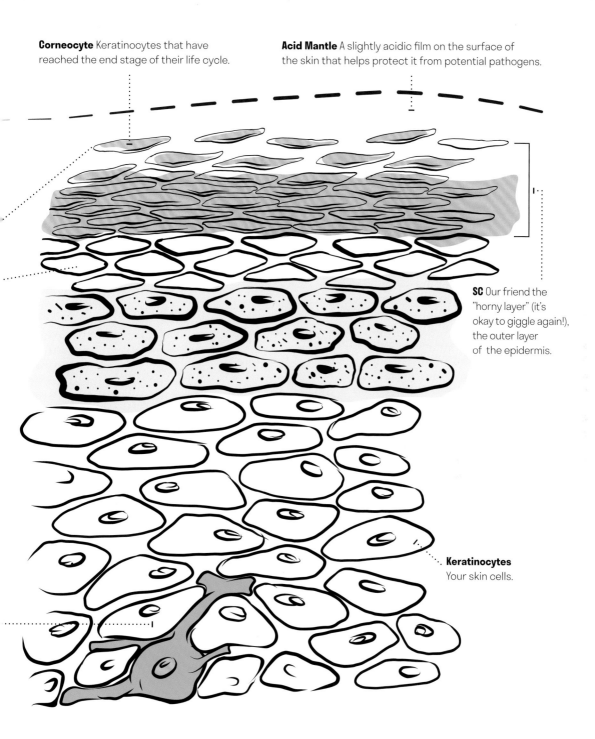

Corneocyte Keratinocytes that have reached the end stage of their life cycle.

Acid Mantle A slightly acidic film on the surface of the skin that helps protect it from potential pathogens.

SC Our friend the "horny layer" (it's okay to giggle again!), the outer layer of the epidermis.

Keratinocytes Your skin cells.

A CLOSER LOOK: DERMIS

Although most skincare excitement happens in the epidermis, the dermis still plays an important role. In fact, it is this skin layer, with all the important structural components, that determines visual characteristics such as suppleness, sagging, and even wrinkle formation. Fun fact! A lot of anti-aging ingredients are tested on cells found in the dermis.

Fibroblasts Fibroblasts giveth, and fibroblasts taketh away. Think of them as the architects of your skin. They are responsible for a wide range of skin functions, including building and destroying collagen, handling inflammation, and even healing wounds.

Collagen & Elastin The structural proteins that hold your skin up. Losing this support over time is what creates fine lines and wrinkles.

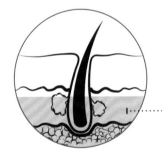

You Are Here

Fat Who could forget fat? It does so much for us, after all! But seriously, your skin depends on that fatty layer, so don't hate too much.

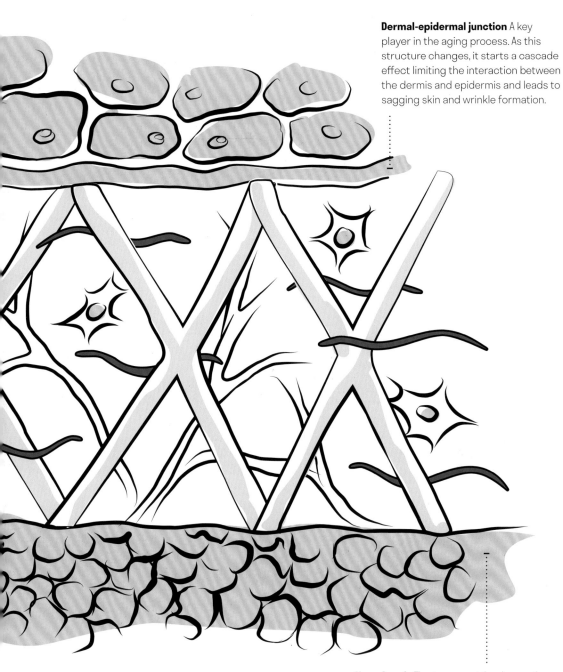

Dermal-epidermal junction A key player in the aging process. As this structure changes, it starts a cascade effect limiting the interaction between the dermis and epidermis and leads to sagging skin and wrinkle formation.

Hypodermis The innermost (or deepest) and thickest layer of skin, also known as the subcutaneous layer or subcutaneous tissue.

SKIN TYPES

The average skin type most likely falls into one of the three main skin types: dry, normal, or oily. However, most people will experience the entire spectrum of skin types in their lifetimes. Skin changes because of changes in lifestyle, climate, stress, hormones, aging—basically, life! While the thought of micromanaging your routine through these changes may sound daunting, you may only need a couple of minor adjustments to your routine to keep up with your skin's changes and get it back on track.

But first, a very important question: What skin type do you have?

Dry Skin (scientific term: xerosis)

Think tightness, flaking, roughness, itchiness, and an overall lack of pliability. It's simply because there's a lack of water in your skin. In science terms, this happens when there's a disruption of the fatty lipid matrix in the stratum corneum—causing water to evaporate out of your skin at a faster-than-normal rate. This kicks off a cascade of unfortunate events. Skin can become thicker (and not in a good, plump, collagen-full kind of way) as the process of sloughing off dead corneocyte cells slows down. This slowdown leads to that undesirable, rough, uneven texture and dull appearance. Unfortunately, skin tends to become dryer and dryer as you age. But not to fear! It's a chemist's job to strategically formulate moisturizers that tackle this very problem. We gotchu, fam!

Key Traits

Characteristics Skin feels tight, dry, and flaky, has a rough texture, and may even feel itchy.

Pros Minimal acne and blackheads, and you can get layer-happy with your routine if you want to.

Cons Flakiness, uneven texture, fine lines, and wrinkles. These issues are often exacerbated by several factors: the climate you live in, plane rides with close to zero moisture, seasonal changes, and aging!

Normal Skin (scientific term: ... skin)

There's a lucky subset of you who have skin that seems just fine: normal, healthy, and hydrated. Skin is bouncy, your skin tone is bright, and you're neither flaky nor oily, because you've got that coveted optimal water concentration in the stratum corneum. Lucky you! That doesn't mean you won't experience unwanted skin changes here and there when life throws new challenges your way. But you already have a fantastic starting baseline! (Jump to our Vitamin C chapter on page 147 to learn more about adding this age prevention active to your routine and keep your skin the subject of envy for life!)

Oily Skin (scientific term: seborrhea)

Does the light shine just right, and you suddenly can't help but feel like you're a disco grease ball? You've got oily skin. This skin type is characterized by excess sebum (aka oil overproduction) due to enlarged sebaceous glands. This is also why acne is often linked to this skin type, since excess sebum leads to clogged pores.

Although you may find excess sebum production kind of gross, sebum actually plays an important part in overall skin health and immunity. Via sebum, the skin is able to produce fat-soluble antioxidants and provide an antimicrobial layer. Another misconception is that having oily skin means that you don't have to use moisturizers. While it's true that sebum can indirectly help SC hydration, sebum production and skin hydration are actually independent of one another. That's because sebum production doesn't involve the lipid matrix, which prevents water evaporation; it involves the sebaceous glands found in the hair follicle. (Refer to the skin diagram to help visualize the difference between the sebum gland and the lipid matrix.) Thus, more sebum output doesn't entirely correlate to a higher ability for the SC to retain water in the skin. Short version? Oily skin can still be dehydrated.

Key Traits

Characteristics Excess sebum, acne, blackhead congestion, large pores, shiny skin by late afternoon, and a general feeling of wanting to wash your face too many times.

Pros Since skin tends to be better hydrated, people with oily skin often have fewer issues with fine lines and wrinkles.

Cons Inconvenient. You're conscientious of how shiny you look for that dinner date. Random breakouts before that speech you have to give. Finding products takes some finagling.

SPECIAL SCENARIOS

In addition to the three most basic types of skin, there are a number of other considerations and conditions you may be dealing with.

Sensitive Skin

To be honest, sensitive skin is not well understood to this day. To keep it simple, sensitive skin here means a skin type that can frequently have reactions to applied skincare products. Sensitive skin is hard for professionals to diagnose and treat, since it's hard to even pin down the root cause in many cases. You could be dealing with a true skin allergy, irritation caused by products, or a unique development over time due to changes in health, lifestyle, and surroundings.

Part of the reason sensitive skin is so hard to manage is that everyone's skin triggers are unique. There are products out there that claim "formulated without skin sensitizers," but the reality is that anything, especially at a high enough concentration, can be sensitizing. Figuring out your skin's quirks and sensitivity triggers can go a long way in building your skincare routine!

Combination Skin

You probably feel like your skin has both oily and dry patches, and just doesn't know what it wants. To be fair, that's actually how skin is. The term "T-zone" was coined for the area with more sebaceous glands than the cheeks-and-chin area.

However, true combination skin is characterized by a much more pronounced difference between the T-zone and U-zone areas. This skin type actually is the most challenging when it comes to moisturization; it typically requires more micromanaging, involving different products for specific skin areas.

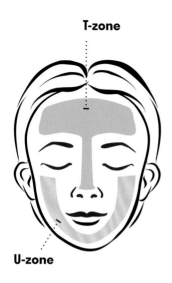

T-zone

U-zone

Key Traits

Characteristics You have both dry patches and an oily T-zone.

Pros You will be the de facto skin guru of your friend group! Combination skin means you'll learn the full range of skincare products out there. (Yeah, we know, it's a cop-out answer . . .)

Cons It is really, really hard to find a one-and-done solution that meets both your dry- and oily-skin needs. Get used to spot-treating dry patches.

Eczema, Psoriasis, and Rosacea

We won't cover these conditions too much in this book, as they really need to be diagnosed and treated by a qualified dermatologist. In a nutshell, all three are common skin conditions characterized by compromised stratum corneum function, which makes it a lot easier for outside allergens to get into your skin and cause irritation. Here are a couple of things we've learned along the way in our chemist careers that might be helpful:

1. **Petrolatum:** Also known as petroleum jelly. It's your friend. It's the gold-standard occlusive that can protect your aggravated skin from the elements.

2. **Skin-friendly pH:** Normal skin pH sits at a slightly acidic ~5.5. Many people with SC conditions like eczema and rosacea have skin pH that skews higher. For people with these conditions, we recommend checking the pH of your cleanser, shampoo, and body wash to help with long-term management.

Ultimately, if you have any of these conditions, it's best to start everyday routine-building with a dermatologist's guidance. In more serious cases, prescription topicals might be necessary. In these cases, think of skincare more as support products rather than key problem-solvers.

PUTTING LOGIC TO THE PRODUCT LANDSCAPE

The main reason you bought this book is probably to figure out the world of products for your skin and its specific quirks. Overall, the skincare landscape seems to have way too many products that all seem to do similar things. Trust us when we say that even we get a little lost, too. Not to fear!

Of all the things you hear about—serums, ampoules, micellar waters, creams, oils, and more—they can all be consolidated by their functions into four simple categories: cleanser, moisturizer, sunscreen, and treatment. You can think of this as a convenient little pyramid. Cleansers, moisturizers, and sun protection set up the base of the pyramid and represent the foundation of good skincare; treatments are the cherry atop the pyramid for that coveted pore-clearing, anti-aging, hyperpigmentation-busting efficacy.

We consider the base of the pyramid the foundation of good skincare. If you want to have a minimalist's routine, cover the three bottom categories.

Cleanser

Every routine should begin with a cleanser to wipe off daily grime and start your routine off on a clean slate. Cleansing can also be considered an age-prevention step, since accumulated grime from the day can have an impact on skin, causing irritation and generating free radicals that can worsen premature aging. Suddenly, the cleanser doesn't seem so basic-biddy, does it?

treatments

moisturizer

fundamentals

cleanser

sunscreen

Moisturizer

Skin's most important job is to serve as a barrier—shielding you from the elements while keeping hydration in. This barrier function, however, becomes compromised when your SC isn't properly hydrated. This is where moisturizers come in! A good moisturizer provides a much-needed dose of hydration while reinforcing your barrier function. This is a super-broad category of products: serums, essences, creams, ampoules, and gel creams are all designed to keep your skin moist and plump. The sheer variety of products makes finding that coveted holy-grail moisturizer quite a challenge!

Sunscreen

The sun is the main external factor that causes premature aging, so this category is extra important. Sunscreens are responsible for shielding skin from sun damage—and, even scarier, skin cancer. Photodamage can cause wrinkles, pigmentation, sagging, texture changes, and a whole host of other issues that are part of premature aging. Do we need to give you any more reasons to use it?

Treatments

If you're covering the three basic pillars of cleansing, moisturizing, and sun protection you already have a solid routine. But if you want to go above and beyond the basics, active ingredients, or actives, are your go-to. These top-shelf products are the glam squad—they usually cost the most and come in pretty packaging. They also basically promise the Fountain of Youth, with ingredients that target and correct skin concerns such as fine lines and wrinkles, tone, texture, and pigmentation.

Actives are a very useful enhancement to a skincare routine, but they can also be confusing. They can have multiple benefits, come in all sorts of product formats, and often require some experimenting so that you can figure out your own regimen and optimal concentration. They're quite a technical and vast category, so half of this book will be dedicated to treatments, and for good reason.

The Chemists' Commandments

The ten rules below are our skincare manifesto, the building blocks of any effective skincare routine. If you read nothing else before you try your first actives or buy into a multistep program, read this. Beautiful skin starts here.

1 **Routine hacking starts with ingredients.** Everyone has their unique skin quirks. Getting comfortable with ingredient lists is your first step in figuring out what works for you—and what doesn't. This is why we started the #decodethatIL hashtag on Instagram (*IL* = ingredient list).

2 **Consistency, consistency, consistency!** The reality is, skincare is a marathon, not a sprint. Consistent good habits will make the difference in the long run.

3 **Overlayering is unnecessary.** Your friend's twelve-step ritual is not inherently more effective than your minimal routine. It all depends on the products, how much you need, consistency, and your own skin's idiosyncrasies.

4 **All good things take time.** We get it—promises of overnight dramatic improvements are alluring. However, any real, visible results take patience and time. Most clinical studies allow at least four weeks for visible signs of skin improvement to occur. For skin concerns such as deep wrinkles or hyperpigmentation, it can take as long as twelve weeks to see significant results.

5 **Concentration is key.** When evaluating a product, look for one that discloses active-ingredient percentages. If the concentration is too low, the product simply won't do anything. If the concentration is too high, you run the risk of irritation. Remember, the dose makes the poison—and the elixir.

6 **You can have too much of a good thing.** It's tempting to have a nightly routine that stacks ten active ingredients to try to get the benefits of all of them. Sadly, adding too many actives just invites more potential for irritation.

7 **Simplicity is key for sensitive skin.** Speaking of less is more—the key to managing sensitive skin is "Easy does it." That ten-step program just contains ten potential sources of irritation, and they all add up! If you have sensitive skin, remember that the simpler your routine is, the more control you have.

8 **Pain isn't gain.** Some painful treatments can be beneficial, but they're rare exceptions and should only be administered by a trustworthy, licensed dermatologist or aesthetician. Your at-home skincare routine should never cause anything more than light stinging. Redness, itchiness, and burning are not signs of your products working—that's just plain old irritation.

9 **Patch test, patch test, patch test!** Even if you don't have particularly sensitive skin, you never know what might cause a reaction. So, test any new product somewhere inconspicuous, like your forearm, before slathering it onto your face!

10 **Skincare is personal. (This is the one that trumps all!)** Last but not least, what works for your friend might not work for you, and vice versa. This is one of the main reasons why skincare is so confusing. Some products that work great for most people can still trigger sensitivity and irritation for some. Don't be too quick to toss something out just because a friend hated it. On the flip side, don't keep trying something irritating because you heard it's an internet cult favorite. Listen to your skin!

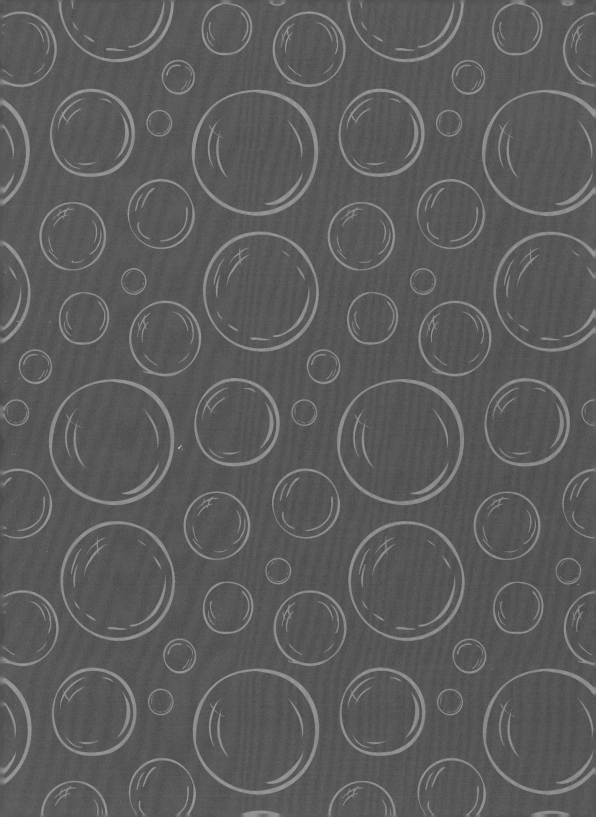

CLEANSERS

Washing your face is perhaps the most seriously underrated part of your skincare regimen. Even if you're among the cleanest of humans, your face still accumulates grime daily. Consider the number of times you touch your face or apply a product in a single day, let alone walk past a running car, a person smoking, a construction site, and so on. Every one of these moments can add microbes, pollutants, dirt, and even stressors day after day. Long-term accumulation doesn't just make you grubby; it will inevitably add to aging. The good news is that all you have to do to counter it is to simply wash your face. Easy-peasy!

WHY CLEANSE

Your face gets dirty. It can't be helped. And not just from the dirt and grime of everyday life. In addition to all that, your skin itself sheds dead cells and secretes sweat and sebum—all part of your skin's natural upkeep—which ensures that it continues to serve as a healthy barrier against outside pathogens. In other words, well, you're shedding, and that's okay! But just know that those dust bunnies that you keep having to clean up around the house? A lot of that is actually you! You can't blame it all on your cat!

Think of washing as an anti-aging step. You may associate washing your face with removing makeup, or maybe our old friend acne prevention—but it's so much more than that. All those particulates can accumulate and cause changes in texture and clarity. They can even add to potential irritation. Pollutants, if left unchecked, can start an unwanted cascade of free radical damage that contributes to aging. You read correctly—your cleanser functions as an age-prevention step.

It should be the first step of your routine. Let's be real: You almost certainly wouldn't be reading this book unless you were planning to invest in your routine. So, let's make sure we're starting off on the right foot—with a clean face! Applying product to an unwashed face can potentially trap dirt and unwanted microbes underneath. In addition to causing breakouts, this can make it harder for your products to penetrate and go to work on your skin. Overall, skipping the cleansing step is pretty counterproductive if you're wanting to build a good routine, so let's start off with a clean slate!

The Takeaway

As unglamorous as cleansers may seem, they do work for you every day. They're the Scottie Pippin of skincare products. Always remember, Michael Jordan might have gotten all the spotlight, but he couldn't have won all those championships without Scottie. That's right, we're saying your cleanser is your routine's Scottie P.

A Very Brief History of Cleansers

Probably one of the oldest forms of skincare, soap, has been around for almost 6,000 years, since the time of ancient Babylon. All sorts of ingredients have been used to make soap, such as ash (lye), animal fat (or pancreases!), plant oils, and more. The first mention of soap—originally used to clean textiles and as a medicine—as a skin cleanser was during the Roman Empire. More commonly, Romans would massage oil into their skin and scrape away the grime—who knew oil cleansers had so much history? Liquid soap didn't come around until the late 1800s, when it was invented by Palmolive. There are many more that we haven't listed here. When you're traveling, try checking out the local soap shops to see different cultures' takes on this skincare essential.

Soaps Around the World

Every ancient civilization had its favorite recipe for soap. You can actually find modern versions of these old-school formulas still in use.

African Black Soap Originating from West Africa, black soap uses the ash from burnt plant bits like bark and cocoa kernels, which is how it gets its signature color. Fans of African black soap swear by its antimicrobial properties for acne-prone skin.

Beldi Soap Also known as Moroccan black soap, *Beldi* has a signature gooey consistency and lack of water in the formula. It's traditionally used with gloves for a deep body exfoliation.

Aleppo Soap Hailing from the city of Aleppo, Syria, this soap is reputed to have historical origins in ancient Egypt and Syria. Its key ingredient is laurel berry oil.

Marseilles Soap Named after the French city of Marseilles, this soap is made with a blend of seawater and olive oil. A very effective cleanser, it's even used for household cleaning—which means it might be too harsh for dry, sensitive skin.

THE SCIENCE OF CLEANSERS

The main "cleansing" components of any cleanser—whether soap, gel, oil, or foam—are its surfactants. These are cool little guys with a hydrophilic (water-loving) head and a hydrophobic (water-hating) tail. The fact that they can interact with both means these guys are a shoo-in for cleansing, since they can pick up dirt and grime, and then be lifted off your face with a splash of water. In a cleanser, these surfactants bunch together to form spherelike micelles. As you apply the cleanser and scrub, the micelles interact with the oil and grime, attach to it, and carry it away with a rinse. That . . . sounds relatively simple, right? Well, yes and no.

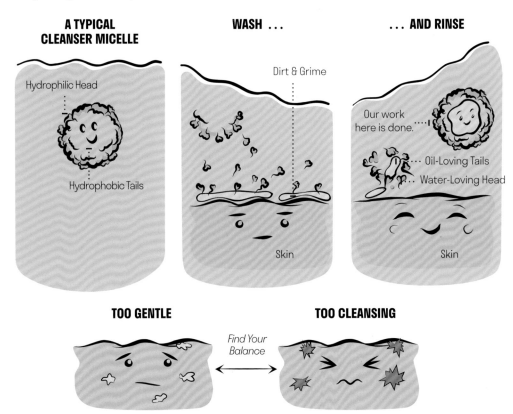

Mr. Micelle vs. Stratum Corneum

While the science is relatively straightforward (we hope!), creating the perfect cleanser is a bit of an art. The ideal cleanser would only interact with the dirt and leave your stratum corneum alone. Unfortunately, some surfactants aren't the sharpest tools in the shed. Some seemingly great surfactants can actually attach to your SC proteins and lipids, causing you skin to have that "stripped" dryness post-cleansing. There are some folks that end up confusing "super clean" as "stripped bare." That's definitely not the goal!

In addition to SC interaction, your cleanser can disrupt your SC's acid mantle. This isn't an actual, physical mantle. The term refers to the lightly acidic environment of your SC and this acidity plays a role in SC's overall health. In fact, higher skin pH has been linked to the disruption of bacteria on skin and is a characteristic of skin problems such as eczema and atopic dermatitis. Unfortunately, many cleansers, including good ol' soap, have very high pH that can be irritating to those with compromised skin.

This is where the chemists come in! Each cleanser you see on the market is the hundredth trial of a chemist carefully balancing stability, support ingredients, cleansing power, irritation potential, and user experience, so read on! We will introduce some key ingredients to look for so you, too, can hack your cleanser game.

I see—so cleansing is all about that delicate balance of gentleness and cleansing power. That squeaky, tight feeling is *not* a good thing!

The Cleanser Landscape

Whether you've used the same brand of soap since high school, or are experimenting with the double-cleansing method, you may be surprised how many cleansing options there are out there.

Powder cleanser Effective, gentle, and adds a bit of physical exfoliation. Just … messy …

Bar soap While some bar soaps can be on the gentle side, most have a higher pH and can be quite stripping.

Gel cleanser Can range from super gentle to powerful acne cleansers that can dry and strip the skin. Read labels carefully!

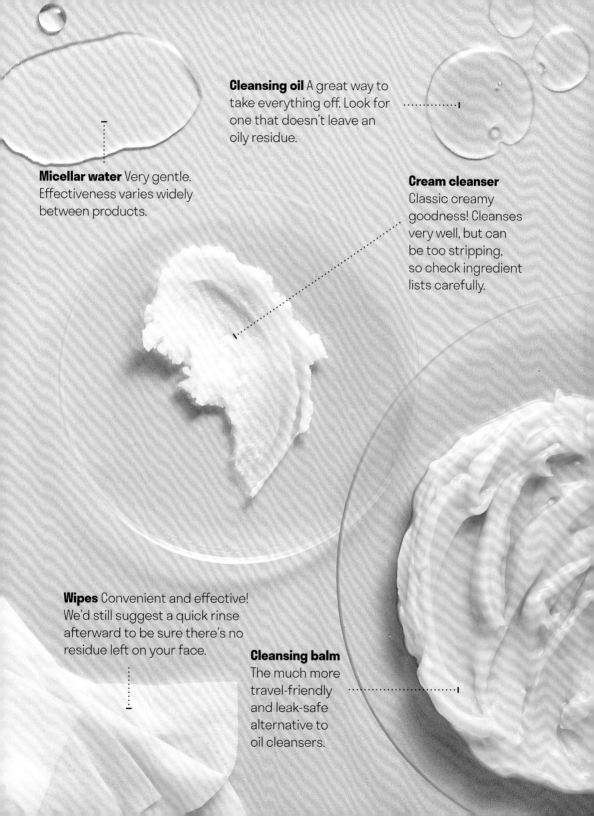

Cleansing oil A great way to take everything off. Look for one that doesn't leave an oily residue.

Micellar water Very gentle. Effectiveness varies widely between products.

Cream cleanser Classic creamy goodness! Cleanses very well, but can be too stripping, so check ingredient lists carefully.

Wipes Convenient and effective! We'd still suggest a quick rinse afterward to be sure there's no residue left on your face.

Cleansing balm The much more travel-friendly and leak-safe alternative to oil cleansers.

IN PRACTICE

Welp! That's a lot of different product types! Just remember, regardless of product format, the trick of the trade is to look for a balance. The right cleanser finds your skin's sweet spot, balancing between effective cleansing and gentleness. So, how does one go about perfecting their cleanser game? That's what we're here for! We recommend that you first familiarize yourself with common surfactants, then check the pH, and finally enhance the cleansing power with support products instead of opting for a more aggressive cleanser.

Why Don't You Love Me?

Meet SLS (sodium lauryl sulfate), one of the surfactants that any cleanser advertised as "gentle, sulfate-free" is going to shun. Now, we just spent an entire section on finding a gentle cleanser, so why are we talking about SLS here? One of the reasons SLS gets blended into a lot of cleansers is because it's cheap and cleans pretty well. It's almost too good at cleansing, hence, this is why it's also used as a positive control in irritation testing. That means that concentrated SLS is actually used to irritate the skin in a controlled manner to test how it recovers with or without products. This is part of the reason SLS gets a bad rep as irritating. But remember the Chemists' Commandment #5: Concentration is key! A small amount blended with other, more gentle surfactants can still make an effective, but mild, cleanser. It's what we do as chemists—we create balanced formulas that bring together the best characteristics of those key ingredients. So, don't freak out too much when you see it in the ingredient list of a cleanser you've been using for the past three years.

Cleanser Tip 1: Know Your Surfactants

For your staple cleansers, gel and cream, products can range from "So gentle—is it really cleansing?" to "Holy moly, my skin actually squeaks!" Other than going off of the often unreliable marketing claims, you should learn to quickly decode that ingredient list! Finding your surfactant BFF can save you a lot of heartache on your cleanser journey. This is why you have us, right? Here's a list of super-common surfactants to get started:

	The Skinny	Who It's For	Who It's *Not* For	How to Spot One
Classic Soap	Great cleansing power, but comes with a very high pH that can be too stripping.	Both normal and oily skin.	Dry, sensitive skin and people with SC conditions such as eczema.	Sodium hydroxide or potassium hydroxide will be high up on the ingredient lists. These are chemicals that turn fatty acids into soap!
Sulfates	Low pH, but still has lots of foam and great cleansing power.	Almost all skin types.	Those sensitized by sulfate surfactants. If you're not sure, look for extra dryness post-cleansing.	SLS (sodium lauryl sulfate) and SLES (sodium laureth sulfate) are the two main surfactants in this category. SLES is the more skin-friendly option of the two.
Coco Betaine	*The* most common gentle surfactant. Skin-friendly, good cleansing power, decent foam.	All skin types looking for a more gentle option.	Ironically, this can be an allergen for some.	The official name of this surfactant is "cocoamidopropyl betaine," but sometimes we chemists get lazy and will call it "coco betaine" for short.
Other Gentle Surfactants	Tried all the most common ones but still haven't found your true love? Try some of the surfactants here.	Those with sensitive skin or chronic dryness.	Those who enjoy a good, luxurious foam. Surfactants in this category are usually low to no foam.	SCI (sodium cocoyl isethionate), glucosides (coco glucoside, lauryl glucosides), cocoamphoacetates, and amino acids are common ingredients in this category. They usually come together in a bundle deal!

Cleanser Tip 2: Consider the pH

If you're sensitive, or have chronically dry skin, psoriasis, or eczema, it never hurts to double-check the pH of your cleanser. Your skin's natural acidity (usually hanging out in the pH 5.5 neighborhood) is an important part of its microbiome and general health. Though scientists don't yet fully understand the complexities of the microbiome and skin pH, try to reach for cleansers with a listed skin-friendly pH (roughly, anything under 6).

 But is pH the be-all and end-all? Actually, healthy skin has the ability to self-regulate pH. So, if you have normal, healthy skin but use a high-pH cleanser that you love, don't freak out! pH is a much more important consideration for those with compromised skin.

CHEMIST CONFESSIONS: **SHOPPING TIP**

EXPENSIVE CLEANSERS

One of our pet peeves is those ridiculously expensive cleansers you see in some department stores. You know the ones: fancy packaging, big price tag, claims to be made with concentrated mermaid tears and promises to wash away your wrinkles and sins. Ridiculous claims are everywhere in skincare, but we find them most eye-roll-worthy in cleansers. Assuming for a second that they did infuse their cleanser with a Fountain of Youth distillate, cleansers are just not really a good way to deliver all that anti-aging goodness into your skin. Remember, a cleanser's job is to take things off, not to add on. Regular-priced cleansers with normal ingredients will do the job just fine.

DECODE That Claim

CLARISONIC CLEANSING

There are a lot of crazy claims out there. Most of you probably already raise your eyebrows at words like "542% more radiant!" or "Turn back time!" But once in a while, there are surprising details and interesting studies behind otherwise nutty-sounding claims. Clarisonic's team did quite a few interesting studies around their brushes. In one, testers applied concentrated amounts of dirt and grime to their face, then compared cleansing efficiency with and without the brush. This study validated that cleansing is much more thorough with a sonic brush. Sadly, Clarisonic is no more, but the test still holds!

Cleanser Tip 3: Boost Your Cleansing Power

So, you have found the perfect staple cleanser that doesn't strip or irritate your skin, but you are just not really confident that you're cleansing thoroughly enough. What can you do to boost cleansing power without irritating skin?

1. **Consider double-cleansing:** Double-cleansing is the method where you start with an oil cleanser, then finish with a gentle cream or gel cleanser. The oil is a cleanser that helps remove all your dirt, grime, and even makeup. The staple cleanser then helps take off any excess grease from the oil cleanser.

2. **Consider a cleansing device:** Cleansing brushes are great additions to your routine that can help your gentle cleanser go further. Sonic brushes are one of the most rigorously tested and validated options you can get—but also the priciest. There are cheaper bristled alternatives, silicone brushes, and non-battery options as well. Just remember, gentle bristles are the way to go!

TROUBLESHOOTING YOUR ROUTINE

Are you still feeling lost? Listed below are some suggested routines for your consideration based on your lifestyle.

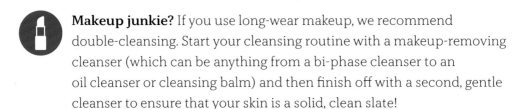

Makeup junkie? If you use long-wear makeup, we recommend double-cleansing. Start your cleansing routine with a makeup-removing cleanser (which can be anything from a bi-phase cleanser to an oil cleanser or cleansing balm) and then finish off with a second, gentle cleanser to ensure that your skin is a solid, clean slate!

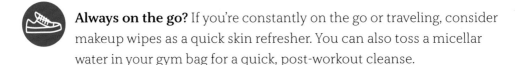

Always on the go? If you're constantly on the go or traveling, consider makeup wipes as a quick skin refresher. You can also toss a micellar water in your gym bag for a quick, post-workout cleanse.

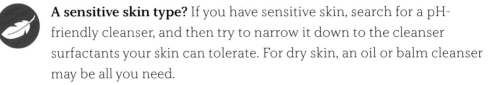

A sensitive skin type? If you have sensitive skin, search for a pH-friendly cleanser, and then try to narrow it down to the cleanser surfactants your skin can tolerate. For dry skin, an oil or balm cleanser may be all you need.

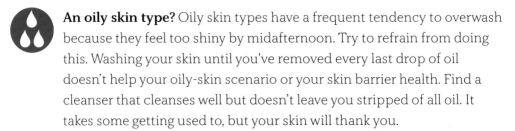

An oily skin type? Oily skin types have a frequent tendency to overwash because they feel too shiny by midafternoon. Try to refrain from doing this. Washing your skin until you've removed every last drop of oil doesn't help your oily-skin scenario or your skin barrier health. Find a cleanser that cleanses well but doesn't leave you stripped of all oil. It takes some getting used to, but your skin will thank you.

To sum up, find your anchor cleanser first. This is a cleanser that covers most of your cleansing needs without stripping. In general, cleansers are pretty personal, so expect some trial and error. But the up-front work will be worth it, because once you find your cleanser, you'll have a fuss-free, reliable staple that's about as second nature as brushing your teeth.

CHEAT SHEET
Cleansers Summarized

Chemist Guidelines:

- Washing your face and removing accumulated grime daily is a fundamental step in your skincare routine, one that actually serves as a long-term anti-aging move.

- A good cleanser is a cleanser that strikes a balance between gentleness and cleansing power.

- Pay for gentle, effective cleansing. Don't pay for fancy anti-aging claims.

Recommended Starting Point by Skin Type:

Oily Look for gel cleansers without that squeaky-clean finish.

Dry Have you met cleansing oil?

Sensitive A safe starting point is a fragrance-free, non-foaming cream cleanser.

Chemist Hacks and Pro Tips:

- Does your skin need a buff? Instead of scrub beads, look at cleansing powders, brushes, and *konjac* sponges for a more gentle but exfoliating experience.

- Try to keep cleansing to max twice a day. That's right: We're talking to you, oily skin types. Washing too often can throw off your skin's moisture needs, and those are important, too.

PERSONAL TAKE

Phew—we're only just beginning our skincare journey, and you already see that there's no one standard answer for every skin type. It's all about your current skin situation and what works for you. That said, we get asked about our personal routines all the time. Here are some things we've learned over the years.

Gloria

Before I started working in skincare and really learning the science behind it, I loooved that squeaky-clean feel of soap-based cleansers. That probably didn't do any favors for my dry skin. When I started transitioning to gentle cleansers, I was skeptical at best, and constantly felt like my skin wasn't really clean. In the long run, though, I really do feel a difference in hydration level. This is especially noticeable in colder months.

When I was a teenager dealing with crazy breakouts, I would wash my face up to four times a day—It scares me thinking about how awful of a routine I had back then. But now, I've gone in the opposite direction and keep it pretty simple. Honestly, sometimes downright lazy—I could be better about eye-makeup remover. In the morning, a splash of water to wake myself up, and a gel cleanser at night. Now I really try to stay away from the cleansers that leave a squeaky-clean finish.

Victoria

CLEANSER FAQs

 Q: How often should I wash my face?

A: At least once a day is solid. We can understand that for some folks, it works to just wake up and splash water on their face, but don't skip the evening.

 Q: Is there such a thing as washing your face too much?

A: Yes, there is. Your skin will give you signs that this is happening: dryness, tightness, and a general dehydrated feeling. Have at most two wash periods a day. (But count double-cleansing as a single wash event.)

 Q: Do I need a special, heavy-duty cleanser if I use sunscreen makeup or long-wear makeup?

A: You actually don't have to use a heavy-duty cleanser with sunscreen and daily makeup. It's the stubborn, long-wear makeup that may need some help from a makeup remover.

 Q: I forgot my cleanser. Is there another alternative I can use?

A: If you've got nothing else on hand, choose a body wash over the hotel bar soap in a pinch. Ultimately though, shampoos and body washes are not formulated with as much emphasis on gentleness. And we don't recommend them as long-term solutions.

 Q: What do you think of people using pure kitchen plant oils like coconut oil or olive oil to wash their face?

A: We get this for the very driest of skin types, which want something that can remove heavy-duty makeup without stripping the skin. But remember Mr. Micelle? Pure oils lack the surfactants that interact with water for a clean rinse; you may end up with more residue than you bargained for. There are oil cleansers that can do more and still won't be drying!

MOISTURIZERS

As the famous saying goes, "Keep skin moist, my friends"—actually, no one says that, but we really can't stress enough how important moisturizing is for skin health. We get it: A super-basic moisturizer just isn't very sexy. But a solid moisturizing strategy is Step One for fortifying your skin barrier so that it can effectively keep the good stuff (water) in and the bad stuff (a whole host of allergens and environmental stressors) out. Without good skin barrier function, all the fancy anti-aging ingredients you splash on your skin are wasted. Let's put the humble moisturizer in the spotlight, shall we?

WHY MOISTURIZE

To understand the long-term importance of moisturizing, we should first revisit some of those Skin Biology 101 terms and concepts. Remember, your skin is like an onion— it's got many layers. The outermost layer is called the stratum corneum, or just SC.

Most skincare products are mainly geared toward treating the topmost layer of the epidermis, the stratum corneum. (If you need a refresher on skin basics, refer to pages 16-21 to get your bearings.)

Skin Barrier Function 1: Keep the Bad Guys Out and Water In

Your stratum cornem is *the* outermost layer of your skin. It's your first line of defense against water loss and outside aggressors. Your moisturizers, in a nutshell, are there to help support your SC in its barrier role.

Skin Barrier Function 2: The Water-Based Stuff

The other component to healthy skin barrier function is natural moisturizing factors (NMFs). NMFs are found in the corneocytes of your stratum corneum and make up about 20–30% of the dry weight of your stratum corneum. These water-based molecules keep the skin elastic and play an integral part in the good-enzyme activity that is key to healthy cell turnover. Common NMFs in your skin are sodium PCA, lactic acid, free amino acids, and urea. Remember: Healthy cell turnover = healthy skin barrier = healthy skin.

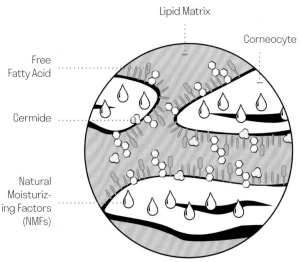

All the Woes of Dehydrated Skin

Your stratum corneum has it pretty rough: The beatings of external allergens, pollutants, UV rays, weather changes, and just plain aging can really take a toll on the SC and prevent it from doing its barrier job properly. When these stressors impact your barrier function, water loss leads to dehydrated, irritated skin—and dry skin has long-term consequences. Neglecting to moisturize leads your poor stratum corneum into a vicious cycle: increasing water loss, letting in outside aggressors, causing inflammation, and leading to more dehydration and less absorption of moisturizers. Uncorrected, this cycle may lead to short-term nuisances like itchy, flaky skin—which turns into fine lines and wrinkles in the long term! This means that a solid moisturizing strategy that evolves with age and the seasons is one of the most important steps in keeping your skin in tip-top shape.

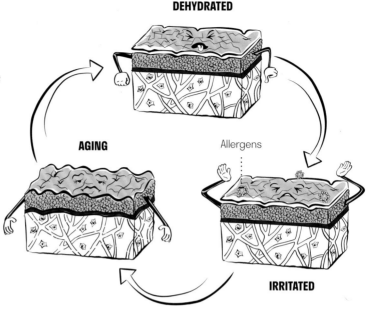

DEHYDRATED

AGING

Allergens

IRRITATED

The Takeaway

As boring as skin moisturizing may sound, without this basic step, the stratum corneum's job gets a lot harder: Layers below the SC are left not as hydrated, our bodies end up less protected from outside toxins, and further skin damage inevitably escalates the aging process. If you don't have time for all the other serums, elixirs, or masques, just grab that jar of lotion and slather away.

THE SCIENCE OF MOISTURIZERS

Even though scoping out a good moisturizer may seem confusing, there's actually some method to the madness. You can divide moisturizers into three categories of ingredients based on their function: humectants, emollients, and occlusives. These three categories work together to fullfill your skin's moisturizing needs. Understanding how these ingredients work, and finding a balance of the categories, is the key to cracking your skin's moisturizing code.

LOTION VS. CREAM

Humectants
Emollients
Occlusives

The reality is that your skin will constantly change. So, instead of hitting the reset button and tossing out your current moisturizer—the one that took you forever to find—figure out which of these categories you might need just a little more of, and you'll master your moisturizing needs through the seasons. Hopefully, thinking about it in this manner will help you confidently troubleshoot and adjust your moisturizing routine when your skin decides to be dramatic.

Category I: Humectants (The Water Stuff)

Humectants are water-grabbing ingredients that help your skin maintain a healthy moisture level, which is key to maintaining that desirable, supple feel. Your skin naturally has its own water-holding system in the form of natural moisturizing factors (NMFs), and the humectants in skincare are there to support them. Some gold-star humectants used in moisturizers are glycerin, hyaluronic acid, and glycols. Humectants are so important that many skincare products are humectant-centric, such as essences, ampoules, hydrating serums, and mists. **Who needs humectants?** From the oiliest to the dryest skin, every type can enjoy the benefits of humectants. They are especially crucial if you want your dry skin to achieve that supple look!

Category 2: Emollients (All Things Oil)

Emollients fill in the rough patches of your skin and instantly give it that soft, smooth feel. These ingredients are usually lighter, oil-based substances such as jojoba oil, caprylic/capric triglyceride, squalane, and coconut alkanes. The most common emollient-centric products are, of course, face oils! These products have become very popular in the past decade. As a result, many brands have released their own "miraculous, antioxidant-rich, wrinkle-correcting, time-stopping, sustainably harvested" face oil. These are great little moisturizer boosters, but don't take the fancy "turn back time" claims too seriously.

Who needs emollients? If you moisturize, there's a solid chance that you already have some emollients in your life. These can be great for both dry and oily skin types; it's just about finding the right oil. Just adding a few drops as a last step to your routine can help elevate your go-to moisturizer on days when it's just not cutting it.

Category 3: Occlusives (Seal That Moisture In!)

From my carcass, I giveth thee petrolatum.

Your skin is your barrier against outside aggressors like UV light and pollutants, but sometimes your skin barrier needs a little help. Occlusives are there to help your skin do that job even better. Occlusives are heavy, fatty, waxy substances that form a physical, water-resistant barrier over your skin to seal in moisture. Some common occlusive ingredients are petrolatum, butters, waxes, and heavier silicones. Occlusive-forward products such as balms or salves are great for spot-treating ultradry patches of skin.

Who needs occlusives? You guessed it! As with the other two categories, everyone could benefit from having some occlusives in their life. But if you're dry, seriously consider occlusives. Slather on that butter, balm, or petrolatum. Trust us—your skin will thank you.

MEET THE INGREDIENTS

If you stalk online skincare-ingredient databases (like we do—we're so cool), you'll see that everything claims to be some sort of moisturizing miracle. But which ingredients are the fairest of them all? Here's a quick introduction to some of our chemists' favorite moist-making ingredients!

Humectants

Nowadays, it seems everything is "hydrating." The reality is, there's nothing out there that ranks all humectants. These water-grabbers can vary a lot in terms of molecular weight and secondary benefits. One strategy is to have a blend of these humectants below to make your hydration game more well-rounded.

Representative Humectant-Forward Products Cosmetic waters, essences, gels, hydrating serums, and most gel creams.

UREA (Molecular weight: 60)

Also part of the NMF category. Unlike other humectants, urea seems to improve barrier function in the long run, on top of instant hydration benefits. In fact, as skin ages, urea becomes more helpful in body care. Yes, there are people out there that claim *"Urea comes from pee! I will not put pee on my face!"* Rest assured that the urea in your cream does not come from urine extract.

LACTIC ACID (Molecular weight: 90)

Also part of the NMF category. Lactic acid is a very versatile ingredient: In addition to being a great hydrator, it can also be used in chemical exfoliation.

GLYCERIN (Molecular weight: 92)

You'll find this in many, many, many products. The reason is—it works! This small molecule can wedge itself into the fatty stratum corneum's lipid matrix, giving skin that nice, supple, flexible texture.

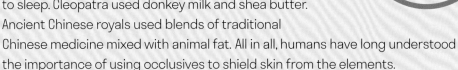

A Very Brief History of Moisturizing

People have been moisturizing for thousands of years! Ancient Roman women applied lanolin before going to sleep. Cleopatra used donkey milk and shea butter. Ancient Chinese royals used blends of traditional Chinese medicine mixed with animal fat. All in all, humans have long understood the importance of using occlusives to shield skin from the elements.

That said, no matter the time period, humans are humans. We get some things right and some things very wrong. The butters and waxes make sense, but there are a few … more questionable stories. Helen of Troy was said to have bathed in vinegar, and Hungarian princess Erszebet (Elizabeth) Báthory allegedly bathed in the blood of young girls. (Don't do that, please. Those methods—particularly the second one—are slightly unnecessary in this day and age.)

PANTHENOL (Molecular weight: 205)

Also known as pro-vitamin B5, panthenol is considered both a humectant and an emollient. It also has great soothing properties.

COLLAGEN (Molecular weight: 300,000)

Many people may associate collagen with anti-aging claims. The reality is, though, that topical collagen can not replace your natural collagen. However, this chubby molecule can be a great hydrator!

HYALURONIC ACID (Molecular weight: anywhere from 20,000 to 2 million+)

Probably one of the most prolific categories of hydrators! You'll hear claims, from hydration to plumping, to even anti-aging. Most HAs on the market are high-molecular-weight polymers—in the 2 million-plus size range. This means molecules sit on the skin's surface, effectively keeping it hydrated all day.

There are other forms of HA that are much smaller. Some studies suggest that smaller HAs can penetrate the skin to plump, hydrate, and even bring anti-aging benefits. However, some people can be sensitized by these small HAs. So, as we always say: Patch test, patch test, patch test (Chemists' Commandment #9).

Emollients and Face Oils

Over the years, face oils have gained a lot of popularity, and every brand is brewing its own version. You'll find that they all seem to be 100% pure, 100% organic, 100% potent, and 100% exaggerated. Regardless of the claims, face oils are a great addition to your routine to boost nourishment, add glow

DECODE That Claim

"X% OF PEOPLE AGREE, SKIN FEELS MOISTURIZED" "STAY MOISTURIZED FOR 24 HOURS!" "SKIN IS 75% MORE HYDRATED!"

Do these common moisturizer claims sound like random numbers yanked out of some marketing exec's behind? It turns out these claims are actually backed by scientific studies. Most moisturizer studies look at two values: hydration and TEWL.

Hydration is measured by a corneometer; this handheld device tracks skin's dielectric (insulative or nonconductive) properties, and reports hydration as a percentage. In a study, you'd typically measure this value on clean skin, moisturized skin, and skin several hours after product application to assess how hydrating a product is.

Remember: *TEWL* (pronounced like *tool*) stands for transepidermal water loss. As the name suggests, it's how much water evaporates out of your skin. The higher the TEWL rating, the worse the skin barrier. This is a really useful way to test several different claims. For example, the tester may purposely damage a patch of skin to create an area with a high TEWL, then apply a particular cream to bring the value back down. Neat, eh?

to dull skin, smooth skin's surface, and improve overall skin pliancy. While oils are lousy stand-alone moisturizers, there are a few key scenarios where a face oil can really elevate your skin routine.

Representative Emollient-Forward Product Face oils

FACE-OIL-TO-THE-RESCUE SCENARIOS

Consider a face oil for the following scenarios:

"I love my moisturizer, but my skin still gets a little dry." Skin constantly changes. Just from stress, age, lifestyle, and hormones, your skin's moisturizing requirements can really fluctuate. If you're loyal to your moisturizer and know your skin needs just a little help in the moisturizing department, consider adding a couple of drops of face oil to give it that oomph.

"I'm dry, but not petrolatum/balm (cactus) dry." We get it—petrolatum and other heavy-duty occlusives feel kind of gross. Not to mention the disco-ball-shiny look is so not in right now. Consider face oils, their lighter (albeit less effective) cousins.

"My skin is a little dull." A quick way to add back that healthy glow to skin is with a couple of drops of face oil.

"I have acne, but my acne topicals are drying me out, and my current moisturizer isn't cutting it." Look for a lightweight emollient such as linoleic acid-forward plant oils or squalane.

New to oils? Consider squalane

Before you try to figure out if you need rose hip, watermelon seed, argan, or whichever plant oil, consider squalane, which is a purely saturated hydrocarbon. In layperson's terms, this means that it's super vanilla and is highly unlikely to irritate you or cause you to break out. It's a great starting point for beginners of all skin types.

Face Oil Shopping Guide

Don't know where to start?! Here are a few popular oils organized with some chemist notes:

 DRY   **OILY**

SWEET ALMOND OIL	**BLACK CURRANT**
You'll find this in everything, even $$$ serums.	**CHIA**
APRICOT KERNEL	**CRANBERRY SEED**
ARGAN	**EVENING PRIMROSE**
AVOCADO	*Easily oxidizes. Use it quickly.*
CAMELLIA	**GRAPE SEED**
Often used in Asian skincare brands.	**HEMP SEED**
JOJOBA	*This is not CBD.*
MACADAMIA NUT	**MARACUJA SEED**
MARULA	*aka passion fruit seed.*
MORINGA	**PRICKLY PEAR**
OLIVE	**ROSE HIP SEED**
SHEA	**SEA BUCKTHORN SEED OIL**
Can come in either butter or oil form.	*Make sure this isn't the fruit oil; it stains.*
	SQUALANE
	Great intro to oils; pretty vanilla. Great for dry skin, too.

 Meet Mr. Vampire. He will remind you to store in his preferred setting—cool, dark, and well sealed. That's right: Oils oxidize and go rancid. Rancid oils change color and smell off. They can end up causing trouble for your skin.

Occlusives

Occlusives are your butters, waxes, and petrolatum. As the name suggests, they help shield your skin from the elements. When your skin barrier function is compromised, occlusives help protect skin and seal in moisture. The catch is that occlusives can feel heavier and greasier than the other categories. People with oily skin need lighter gel creams that contain little to no occlusives, while those with dry skin may want to opt for a heavier cream that has more. We do always recommend having an occlusive balm (think Vaseline, Aquaphor, and the like) to spot-treat those troubled areas!

 Representative Occlusive-Forward Products Balms, butters, salves

PETROLATUM (good ol' Vaseline)

Still the gold-standard occlusive. In recent years, it's seen some terrible press in chemophobic groups along the lines of "It causes breakouts and also cancer." The reality is, cosmetic-grade petrolatum is highly refined—which means it's exceptionally "clean" (free of potentially irritating and harmful residue). It doesn't clog pores, but it *can* cause breakouts by sealing in dirt and grime if you are lazy about cleansing.

MINERAL OIL

The lighter version of petrolatum is much less occlusive, but feels better in a cream formula. Fun fact! You'll be surprised how much these ingredients affect the feel of the final product. Just a 1% difference in mineral oil or petroleum jelly content completely changes the feel of the cream.

SHEA BUTTER

Shea in butter form is a great, versatile natural occlusive. With the clean movement, you may stumble upon "raw shea butter." People think *ultranatural* means ultrasafe and free from scary chemicals. The reality is that plants are complex. In its raw form, shea butter can be irritating. We recommend sticking to refined shea; you'll have a range of options for texture.

LANOLIN

This is a great occlusive derived from wool. It can be an allergen, though, so look for medical-grade lanolin and definitely make sure you patch-test before using!

A note about coconut oil

You'll find a lot of coconut-derived stuff, but not all are created equal. For example, caprylic/capric triglyceride are light esters derived from coconut oil. These light emollients are generally friendly to oily skin. Coconut oil—the stuff that a lot of DIY skincare recipes will use, and the same stuff you use to make coconut shrimp—is much heavier and isn't very friendly to oily, acne-prone skin. You also end up smelling like . . . coconut shrimp.

WAXES

More great occlusives that often come from natural sources such as beeswax, candelilla, carnauba, and others. Their main drawback is that they're, well, waxy. And wax's high melting point makes it unsuitable for high-level use.

SILICONES

This is a vast category, but regardless of the material type, the general unifying theme of silicones is simply, "These feel awesome!" Silicones can be effective occlusives without much of the heaviness. If you have oily skin, silicone gel creams can be a great way for you to boost your hydration game between seasons—or when your skin goes through changes—without that dreaded greasy feel. How do you spot a silicone gel cream? Look for terms such as dimethicone or "dimethicone crosspolymer" fairly high on the ingredient list.

Some believe "it's like a plastic bag that just doesn't let my face breathe!"

It's not completely clear how this rumor came about, but silicones are an incredibly broad category. They can range from light oils to thick waxes, and can even be functionalized to emulsify creams. While you may come across an article that's wildly critical of silicones, it's important to try silicones out for yourself to really know if your skin likes them or not. Silicones can create beautiful textures. They can even mattify and even out skin.

What about Other Body Parts?

It seems like there are more and more products out there made for very specific areas of skin—eye, lip, neck, left butt cheek—but is there really a need for all the different products? Here are some tips about specific areas:

The skin is thicker here, so if you're struggling with dry hands and feet, look to occlusive balms to prevent cracked skin. You've been warned: Textures get greasy, so grip will be compromised.

The eye area is more delicate and has a thinner SC—but basically needs the same ingredients as the rest of the skin on your face. There's nothing wrong with extending your facial products into the eye area as long as they don't irritate your eyes. We're only talking about moisturizing needs here. Turn to pages 222-229 for anti-aging eye care.

Lips could really use their own products. Not only is lip skin found to have a thinner SC, but it's also completely devoid of sebaceous glands and skin pigments (melanin). This is why you need heavy-duty occlusives such as petrolatum, lanolin, and beeswax to help shield your delicate lip skin from the elements. Without melanin, your lip skin is also susceptible to UV damage. Consider a lip balm with SPF during the day!

Treat that precious left butt cheek like the rest of your body—with a good ol' body lotion. We've recently seen luxury body lotions pop up, but you can save your money here. Body skin is relatively resilient and durable, so a simple drugstore lotion will suffice. As you get older and your skin cell turnover slows down, dry skin will become a bigger issue, so look for body lotions with lactic acid and urea to help the skin barrier.

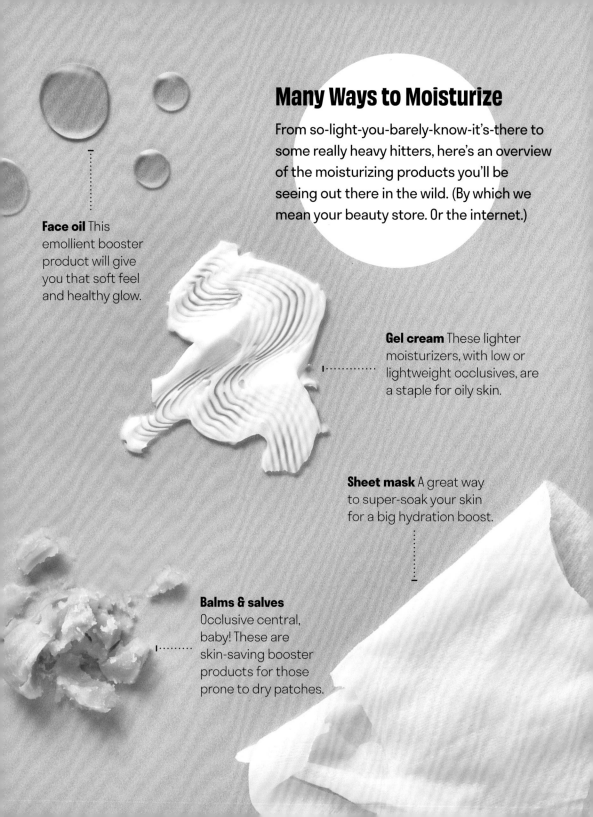

Many Ways to Moisturize

From so-light-you-barely-know-it's-there to some really heavy hitters, here's an overview of the moisturizing products you'll be seeing out there in the wild. (By which we mean your beauty store. Or the internet.)

Face oil This emollient booster product will give you that soft feel and healthy glow.

Gel cream These lighter moisturizers, with low or lightweight occlusives, are a staple for oily skin.

Sheet mask A great way to super-soak your skin for a big hydration boost.

Balms & salves Occlusive central, baby! These are skin-saving booster products for those prone to dry patches.

Creams These heavier moisturizers really bring it for dry skin, with higher doses of occlusives, including petrolatum or butters.

Mists & toners A simple way to add a smaller, less concentrated amount of hydration.

Serums & gels A great way to add a real concentrated boost of hydration.

THE PRODUCT LANDSCAPE

If you walk down the shopping aisles hunting for a moisturizer, you'll quickly find that the landscape is vast and confusing. But you can actually break it down into your staple moisturizer products and your booster products.

STAPLE PRODUCTS

	Lotions	Gel Creams	Creams	"Waters"
aka	Simple moisturizers	Many "oil-free" moisturizers are gel creams.	Traditional moisturizers in a jar. Trendy "night masks" are usually creams too.	Toners, micellars, mists, essences
H/E/O Ratio*				
Who/What/ When	This is the most generic, basic moisturizer. Perfect daily upkeep staple for lost souls and normal skin.	Typically humectant-heavy. If it includes occlusives, it usually uses silicones or other lightweight options.	More heavy-duty than your basic lotion. The increased occlusive content makes it perfect for those with dry skin.	First step post cleansing. This is more of a feel-good than a must-have for most skin types.

* This table is a general breakdown. Different brands may interpret these terms differently. H/E/O is referring to the humectant, emollient, occlusive ratio.

Putting It All Together

Here it is—your chemist-approved starting point for moisturizers. It can be a bit of a Goldilocks experience to try to find your perfect moisturizer, but once you do, you can move on to the booster category to tailor your moisturizing to your ever-changing skin needs. And, as always, you know your skin better than we do! If you have a go-to that doesn't quite fit our assumptions but makes your skin look and feel awesome, don't let anyone talk you out of it. Including us!

Staple Products vs. Booster Products Staples are your more traditional moisturizers, which are a combination of humectants, emollients, and occlusives in one milky, creamy formula. Booster products are usually made up of only one type of moisturizing ingredient. They are great to have in your arsenal so you can adjust to your skin's needs when you find your staple product is not quite cutting it.

BOOSTER PRODUCTS

Gels	Masks	Oils	Balms
Hydrating gels, serums, ampoules	Most sheet masks	Face oils	Salves, butters
A seriously hydrating gel can be enough for the oiliest skin. This is also a great booster for those with dry skin.	A hydrating "pamper me" moment suitable for all skin types.	Last step or mixed with your staple product. Depending on the type of oil, it could be a great booster for all skin types.	Last step. Great spot treatment for dry or irritated patches.

Am I . . . Normal!?!	I'm. So. Dry.	I'm a Little Oily	Feeling Sensitive
Gel Cream or Lotion	Gel and Cream and/or Balm	Gel or Gel Cream	Petrolatum Balm
You're blessed! Tight and dry skin is just a myth or a winter special. Just a simple gel cream or lotion should suffice!	Layering is your friend! Add a balm to target extra-troublesome, flaky patches.	Gels might be all you need to avoid midday shine. Simple gel creams are perfect for mildly oily or combination skin.	Look for skin barrier–reinforcing ceramides, cholesterol, fatty acids, or soothing ingredients.

TROUBLESHOOTING YOUR ROUTINE

Based on the charts of all the crazy product types, moisturizers are a massive category. Not to mention that many of these product types have overlapping functions. So, when you're developing a good moisturizing routine, try following these simple guidelines according to your skin type.

Oily Skin

For those of you with oily skin, lucky you! That usually means your skin is already better at sealing moisture in. Your moisturizing routine can be a lot simpler than it is for those with dry skin. The key is to focus on a humectant-centric routine with a dash of lightweight, oily-skin-friendly oils. See pages 54–58 for a quick recap on how to choose a face oil! Another great one-and-done product choice is silicone creams. Look for products with dimethicone or dimethicone crosspolymer high on the ingredient list for weightless moisture.

Dry Skin

Layering is your friend! We recommend having an arsenal of products to take you through various skin changes.

1. **Humectants:** Start with a loaded hydrating serum. Although ingredients such as hyaluronic acid are super popular, we recommend serums that feature a blend of humectants over a serum that only uses one star ingredient.

2. **Moisturizer:** Look for lotions or creams that have an occlusive such as mineral oil, petrolatum, or shea butter relatively high up the list (think third to fifth position).

3. **Balm/petroleum jelly:** Keep one handy for those pesky little dry spots that may pop up here or there.

Sensitive Skin

Rule 1: Keep it as simple as possible and start with one single moisturizer. More steps = more chances of irritation.

Rule 2: Seek out helpful skin barrier–supporting ingredients. Lipid-mimicking ingredients like ceramides, cholesterol, and fatty acids can really help elevate your moisturizer game and get your skin back on track. Ceramides in particular can help strengthen your skin barrier, help it retain moisture, and even protect it from outside environmental aggressors. Studies have even shown that ceramides are helpful to those who struggle with eczema and psoriasis. (Ask your dermatologist how to incorporate this ingredient into your treatment plan.) It's also a great idea to include soothing ingredients into your moisturizing routine. Check out pages 175–178 for some of our favorite soothers!

Can You Make It Even Simpler?

Here's the absolute minimum you need to know when troubleshooting your moisturizer routine:

1. **If you want to add a booster products, pick at most two products in the toner, essence, water, serum, and ampoule category:** These products are mostly humectant-only. The more cumbersome your routine becomes, the less likely you are to be consistent. Stick to just one or two products from this category.

2. **Don't get greedy:** We all want that one product that solves all our problems. But sometimes that means you get a Franken-product that doesn't let you benefit from any of the good ingredients. So, as tempted as you may be to buy a serum that brightens, anti-ages, hydrates, *and* battles pigmentation changes, find a simple yet effective hydration serum first. Remember: hydrated, healthy skin first, and then move on to the other good stuff.

3. **Remember layer order:** If you end up lost and confused in your layers, just remember to start from most water to least water.

TROUBLESHOOTING SEASONAL CHANGE

Environment can be a big factor in your moisturizer needs. You may find that moving or vacationing to a new, unusual location suddenly throws your moisturizing needs out of whack. Some things to consider include how hot (or cold) and how moist the air is going to be, both indoors and out. After all, your dry, chapped winter skin isn't going to get much happier when you head inside to an overheated ski cabin!

 Headed to the frigid Arctic? Be sure to have an added oil or balm on hand to help with dry, chapped skin.

 Vacationing in the tropics? Consider a more weightless hydrating gel or lighter moisturizer that's able to keep up with your skin's moisturizing needs without feeling too heavy in the humid heat.

 That meditation retreat in the desert? Either add a hydrating gel underneath your current moisturizer or bring a lightweight oil to add over it, to minimize that tight, dry, parched feel.

 Wow! Skin is so dynamic! I finally understand why my skin gets so lizardy on the family ski trip! This year, I'll definitely pack a balm to fight that scaly look and feel!

CHEAT SHEET
Moisturizers Summarized

Chemist Guidelines:

- Dehydrated skin causes a cascade effect of skin issues, leaving it vulnerable to inflammation and wrinkles.

- Skin is dynamic, so it's important to find a balance of humectants, emollients, and occlusives that will solve your skin's moisturizing needs.

- Tackle skin hydration first before getting into heavy-duty actives.

Recommended Starting Point by Skin Type:

Oily Water gels and gel creams. Dimethicones can be helpful in keeping things light.

Dry Thicker lotions and creams. Have a face oil or balm on hand for extra help.

Sensitive Try a fragrance-free lotion with a simple ingredient list.

Chemist Hacks and Pro Tips:

- **Winter is coming:** When seasons change and skin changes, see if you can add a booster product (water gel or face oil) instead of throwing out your moisturizer and starting all over again.

- **The dreaded airplane**: Focus on occlusive balms. All you need is to prevent water from leaving your skin, so seal that moisture in with a good balm.

PERSONAL TAKE

By now, you can see that there's no one standard answer to skincare. It's all about your current skin situation and what works for you. Here are some of our personal notes on this category.

Gloria

My skin swings from normal/dryish to "OMG it's PARCHED" between seasons. Hydrating serum is a staple I have throughout the seasons. I prefer using just one highly concentrated hydrating serum that has a good blend of different water-grabbers, instead of layering 574,289 hydrating products. When my skin gets extra dry, a petrolatum-based balm becomes my best friend. It looks and feels kind of gross, but lathering up my whole face in petrolatum in desperate times really helps.

With combination skin, this category can be very infuriating to explore. In fact, I usually end up very skeptical of "combination" products, because they're often still too heavy, or end up being loaded with mattifying powders that don't really help skin in the long run. After a lot of painful trial and error, my anchor product is a light lotion that handles my face at its oiliest moments. I'll layer in a hydrating serum (heavy on the humectants) on drier days. Ultimately, being able to fine-tune my moisturizing routine has been a lot more successful than leaving it up to a single "combination" moisturizer that ends up just doing a mediocre job.

Victoria

MOISTURIZER FAQs

1

Q: I heard using hyaluronic acid in dry climates ends up making your skin drier.

A: If you are solely relying on hyaluronic acid, it can certainly be a problem. Use some occlusives to seal in moisture in a dry climate!

2

Q: If I use too much chapstick, will my lips forget to hydrate?

A: Sadly, your lips are already a little handicapped when it comes to staying hydrated, since the lips don't have any sebaceous glands. There's nothing wrong with using chapstick often, but if you're experiencing long-term dry lips where you're having to apply multiple times a day, there may be a bigger issue here. Make sure you're using proper lip balms that are heavy in petrolatum and butters and have fewer cooling ingredients, such as menthols and peppermint oil.

3

Q: I have oily and/or acne-prone skin—will moisturizers with oils make me break out?

A: This is highly product-dependent! In fact, the "oil-free" claim is probably one of the most pointless claims in skincare. The reality is, finding your pore-clogging triggers could take a bit of detective work. Always patch-test a product to figure out your skin's quirks!

4

Q: Does everyone need moisturizer? Even if I've never used one my whole life?

A: If your skin has been doing perfectly fine without a moisturizer, then please donate your DNA to science so we can figure out the secret genetic code behind perfect skin. In all seriousness, though: No, you don't need a moisturizer if your skin does fine without one. However, if you're noticing unwelcome changes and want to start picking up some products, moisturizer is an important foundation of virtually every rounded-out routine.

SUN PROTECTION

The sun is *the* science-backed skin-aging culprit. In fact, it is such a significant factor that this type of aging is called photoaging. It has been extensively linked to premature fine lines and wrinkles, texture changes, leathery appearance, pigmentation, and downright DNA damage. And yet, for some of us (we're included), we can't help but want that golden, healthy tan. It doesn't help that sunscreen isn't the prettiest of formulas—gooey, white, heavy, greasy, tacky, and it's even got that signature scent. Applying this on the beach makes you think more of that uncle with the shiny white nose who always reeks of sunscreen than the glamorous Miami Beach life. But it's just a matter of finding your perfect sunscreen match. In the long run, better sunscreen habits lead to better skin. It's the true anti-ager! Here, we'll make sure you're getting proper protection with a formula you love. Let's rekindle your love for sunscreens.

WHY SUNSCREEN?

To protect yourself from skin cancer. Thanks for reading! ... Okay, just kidding. For even the laziest of skincare users out there, if there's just one thing you should do, it's sunscreen. Yes, more than moisturizing. More than cleansing. Protect thy skin from the angry rays of the sun, people! This is truly the cornerstone of good skincare.

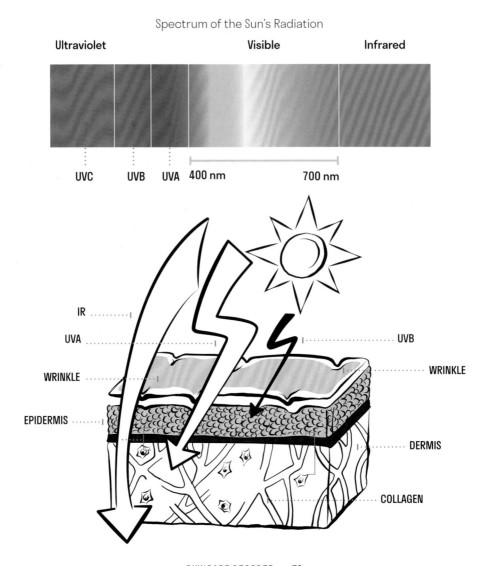

Spectrum of the Sun's Radiation

Ultraviolet   Visible   Infrared

UVC   UVB   UVA   400 nm   700 nm

IR

UVA   UVB

WRINKLE   WRINKLE

EPIDERMIS

                                             DERMIS

                                             COLLAGEN

The Sun Emits Two UV Rays Responsible for Skin Damage

Let's all remember that the sun is a giant nuclear reactor that emits energy across a wide spectrum, including visible light, infrared (heat), and ultraviolet. Among the three types, ultraviolet is the main type of radiation responsible for causing skin damage, ultimately resulting in premature photoaging.

The sun emits three types of UV rays. It's helpful to know the difference, as products will often refer to the different types.

UVC (200-280nm) The shortest of the three wavelengths, therefore the most powerful. Luckily, this one gets filtered out by the atmosphere, and, thus, we can rest easy and ignore it. (Yay!)

UVB (280-315nm) The mid-length wavelength, UVB makes up 5% of the UV that reaches our skin. Because of a shorter wavelength in comparison to UVA, this only reaches the epidermis. UVB is responsible for sunburns and delayed tanning, and it is the main culprit of skin cancer. SPF value on a product denotes UVB protection.

UVA (315-400nm) The longest of the three wavelengths, UVA makes up 95% of the UV that reaches our skin. UVA can reach all the way down to the dermis and is responsible for deep photoaging and enhanced skin-cancer development. We always remember the two as *B* is for *Burns*, *A* is for *Aging*.

But my vitamin D! Is sun exposure necessary to get it?

It's true that there is a growing deficiency in vitamin D worldwide, but with a modern problem comes a modern solution. Instead of relying on the sun to provide energy for our bodies to synthesize vitamin D, consider increasing your intake of foods high in vitamin D: fish, eggs, and milk. Even adding a vitamin D supplement is sufficient. We like this route better than trying to calculate how much sun you need before you're getting into skin-damaging territory. Why risk it?

Sun Scenarios

This may be a bit of a no-brainer, but location does matter when it comes to sun exposure, so let's look at a few different "sun-arios"!

 On a mountain Higher altitude means thinner atmosphere, which also means more UV exposure. Definitely apply sunscreen when you're hitting the slopes.

 On a plane At very high altitudes, the UV rays will definitely be stronger. Do you need to worry about exposure if you have a window seat on a plane? Well, plane windows are made of plexiglass, and while they do block some UV, there's still some UVA radiation that's able to filter through. What you can do is close that shade, put on an eye mask, take a big hit of melatonin, moisturize, wrap yourself up in full baby-burrito style, drink water, and get some rest!

 Overcast This is the term for when clouds cover 95% of the sky, which means you do have less sun exposure. But even if visible and infrared light is decreased, know that UV rays are still touching your skin. We'd say it's still a good idea to use SPF, especially if you're out and about.

 Partly cloudy Partly cloudy situations are funky because of the broken-cloud effect. Sometimes, when the clouds are situated just right or are scattered but still dense enough, they can actually scatter UV and cause the effects of the rays to be as much as 25% stronger.

 Driving or working indoors The glass in a car or in a window at your office or your home blocks UVB, but not UVA, so you're still exposed to photoaging.

A Brief History of Sunscreen

Historically, suncare awareness developed more out of aesthetic desires than actual sunburn protection. Many cultures shared the same desires for lighter skin. The Egyptians used natural ingredients like rice bran, jasmine, and lupine to prevent tanning. The Greeks used olive oil as sun protection. In Myanmar, they use *thanaka*, a cream made of ground-up tree bark, to smooth skin, prevent acne, and prevent sunburn. Up until the early 1900s, many pharmacists would create their own sunscreen concoctions using olive oil and almond oil.

By the 20th century, sunscreens became commodified and incorporated actual sunscreen filters like zinc oxide and benzyl salicylate. The French chemist Eugène Schueller, the founder of L'Oreal, was the first to produce a true chemical sunscreen. We must also give credit to Austrian chemist Franz Greiter for not only developing the first truly effective sunscreen formula but also creating the SPF (sun protection factor) metric that we still use today. Other than topical sun protection, people have also been donning sun-protection gear throughout history. Based on historical descriptions and images of items ranging from sunbrellas and bonnets to terrifying leather masks called visards, it seems we've known about the dangers of sun damage for a long time!

BONNET

VISARD

SUNBRELLA

So, that's thorough sun-protection gear? . . . I think I'll just wear sunscreen!

The Leading Cause of Skin Cancer Is Too Much Sun Exposure

UVB is the main culprit that causes sunburns and, eventually, skin cancer. Its short wavelength means it packs enough energy to directly damage your skin cells and even DNA. This is why if you get a sunburn more than five times, your melanoma risk doubles.

Other than direct damage, UVB and UVA can also cause damage by triggering free-radical development. The dreaded free radicals are reactive molecules that unselectively attack your skin cells, proteins, and even DNA, causing long-term damage.

UV RAYS GENERATE FREE RADICALS

" Wait! My antioxidant serum tells me it can prevent sun damage. Does this mean antioxidants are sunscreens? "

Well, no, but think of it this way: Sunscreens prevent the development of free radicals, while antioxidants quench the developed free radicals. Thus, sunscreens and antioxidants are a lovely pairing. For more on great antioxidants in skincare, check out pages 148-151.

Too Much Sun Exposure Is *the* #1 External Factor That Causes Premature Aging

Aside from skin cancer risks, sun exposure is also the major culprit of skin aging. Ultraviolet damage causes everything that you don't want: sagging, rough texture, leatherlike thickness, hyperpigmentation, wrinkles, dullness, and more. As the sun causes damage to the skin, your own fibroblasts are triggered as an inflammatory response and will start to break down your collagen. As another defense mechanism, your skin also thickens, and hyperpigmentation pops up from an overproduction of melanin packets. Dry, dull, rough textures and wrinkles develop from long-term damage. Even that golden, healthy tan is a sign that your skin is defending itself against sun damage. In fact, UV damage is such a significant factor in premature skin aging that there's a term dedicated to it: photoaging.

Too much UV exposure triggers fibroblasts to break down collagen.
* Fibroblasts don't actually use forks and knives.

Hopefully, we haven't scared you into retreating underground. Ultimately, we just want to emphasize good sunscreen habits. Sunscreen is a key player in a solid routine and will save you a lot of money on products and procedures in the long run. In fact, that's why we consider sunscreen the ultimate anti-aging product!

THE SCIENCE OF SUNSCREEN

Sunscreen protects your skin by absorbing harmful UVA and UVB rays so your skin doesn't have to. While many believe that sunblock actually physically blocks the skin from the sun, instead, it generally absorbs the UV rays and dispels the resultant small amount of energy as heat. It's important to remember that because sunscreens protect you from skin cancer, they are heavily monitored and regulated by the FDA. In fact, this is why you'll see sunscreen labels that have a drug facts section so you won't have to search for the active ingredients from a sea of ingredients.

When you're hunting for your sunscreen love, there are three important factors to check off: UVB protection (SPF), UVA protection, and texture.

UVB Protection, AKA Sun-Protection Factor (SPF)

First, SPF is not on a linear scale. It might be tempting to think an SPF 30 sunscreen must offer twice the protection of an SPF 15, but that's not how SPF value works. SPF 15 actually means that with proper application, it only lets

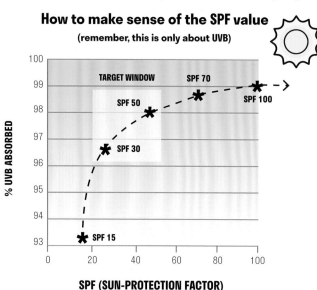

How to make sense of the SPF value
(remember, this is only about UVB)

TARGET WINDOW   SPF 70

SPF 50   SPF 100

SPF 30

SPF 15

% UVB ABSORBED

SPF (SUN-PROTECTION FACTOR)

$1/15$th, or about 7%, of UVB through. This means that an SPF 15 sunscreen blocks roughly 93% of UVB, while an SPF 30 sunscreen blocks about 97% of UVB. Definitely not "twice the protection." Fun fact! This is why the claim "SPF 100" is banned in the EU, for the misleading insinuation that it can block 100% of UV light. What does this mean practically?

For daily use, we recommend choosing a product between SPF 30 and SPF 50 to hit that sweet spot that balances between sufficient protection and a pleasant texture.

UVA Protection

SPF only covers part of the equation, since it addresses just the UVB portion of the rays. You also need proper protection from longer-wavelength UVA rays; these don't burn skin, but actually penetrate deeper into the skin and are the main culprits behind photoaging. In the US, look for the words "broad spectrum" on your sunscreen to make sure it's tested for UVA protection as well.

SPF is fairly universal, but you'll see different labels for UVA protection around the world. Japan uses a PA system, ranking UVA protection from PA+ up to PA++++. This system is also used in other countries worldwide. In the EU, to get a UVA label, the sunscreen has to have around the same protection level as a PA+++ sunscreen. In a nutshell, this is yet another reason why US sunscreens are a bit behind the times.

DEBUNKED:
All the Other Rays

UVA and UVB are not the only wavelengths from the sun. We have noticed a trend of products claiming protection from rays from a full spectrum. Unlike UVA and UVB protection, these claims are NOT regulated and testing is not required. Here's a quick breakdown about why you shouldn't worry too much about these rays and should take the claims with a whole teaspoon of salt.

- **UVC Protection** UVC has even higher energy than UVB. That must mean even more intense sunburns, right? Actually, UVC gets absorbed by the stratosphere and ozone, so ignore this claim.

- **Blue Light** Do I need to protect myself from my phone and computer screens? To put your mind at ease, the amount of blue light you get from the computer is substantially less than from the sun. Most of the ingredients claiming to protect you from blue light are antioxidants … so there's nothing majorly new. We wouldn't buy a product specifically to protect from blue light.

- **Infrared (IR)** Infrared does penetrate much deeper into the skin than even UVA. There are actually some studies that show how this can be damaging to skin in the long run. The solution? Antioxidants to combat any potential free-radical damages. If this sounds like a bunch of mumbo jumbo, check out pages 148–159 to learn how to choose a good antioxidant for your routine!

Above All Else, Texture Is King

As chemists, we want you to know that formulating sunscreens is a royal pain in the butt. They throw every challenge at formulators, and it really shows when you try to use them. Sunscreens can often end up feeling greasy and heavy and make you look mime-worthy, with a white-cast finish. Despite all the benefits in protecting your skin, sunscreen isn't an easy product to incorporate into your daily routine. Hence, that is why we want to emphasize that sunscreen texture is king. Remember: For the entire body, you'll need about a shot glass's worth of product, and for the face you'll need about ¼ teaspoon (roughly the size of a quarter) to get proper sun protection. So, keeping that dosage in mind, the more powerful sunscreen will be the one you're willing to use every day and happily reapply every two hours. So, don't spend the time, effort, and emotional despair fighting with yourself to use an absurdly high SPF. The good thing is, gone are the days that you have to use that sunscreen goop—there are options! So, let's go through the formats and find your sunscreen love!

CHEMIST CONFESSIONS: SHOPPING TIP

CORAL REEFS AND SUNSCREEN

Chemical sunscreen filters oxybenzone and octinoxate are banned in Hawaii due to their impact on the coral reef. However, many other sunscreen filters have also fallen under scrutiny for their coral reef safety.

   To put it into perspective, the impact of these filters is still relatively minor compared to much bigger culprits such as global warming. Unfortunately, shopping for "reef safe" doesn't mean a whole lot since this is not a regulated claim. For now, you can reach for water-resistant mineral sunscreens for peace of mind . . . or until we obtain more learnings on this issue.

Get to Know Your Sunscreen

Broad Spectrum: Denotes UVA protection. In the US, a simple "Broad spectrum" label is used. For EU sunscreens, look for a "UVA" label. In Asia, look for a "PA+" label.

Where you'll find sunscreen filters listed.

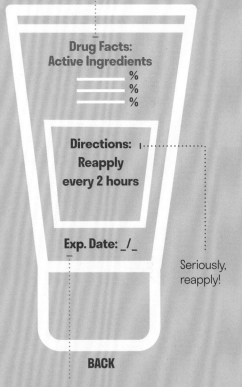

Broad Spectrum

SPF #

FRONT

Drug Facts:
Active Ingredients
%
%
%

Directions:
Reapply
every 2 hours

Exp. Date: _/_

Seriously, reapply!

BACK

Water Resistance: Skin is submerged in water for the allotted time, then tested for protection. In the US, SPF rating is only tested for 40 and 80 minutes. Sunscreen with water-resistance testing is also recommended for sweaty activities. Sweatproof and waterproof are not legal claims.

Expiration Date: Not all bottles have this, so Sharpie it on to keep track and toss the bottle when it's time to go.

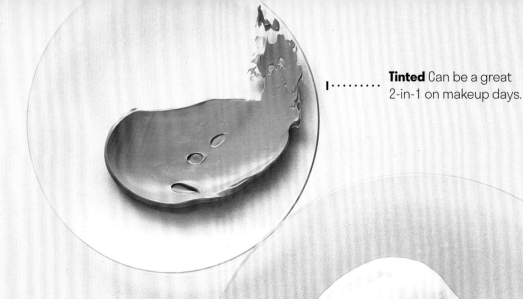

Tinted Can be a great 2-in-1 on makeup days.

Chemist Guide to Finding Your Sunscreen

Sunscreen is the last step in the routine. No matter your preferred formula type, the name of the game is to pick something you love and can layer on thick!

Spray Convenient, easy to reapply, but can be drying. Best to spray into your hand, then rub on for coverage.

Mousse Newer sunscreen format that promotes applying the proper amount because of its fun whipped texture.

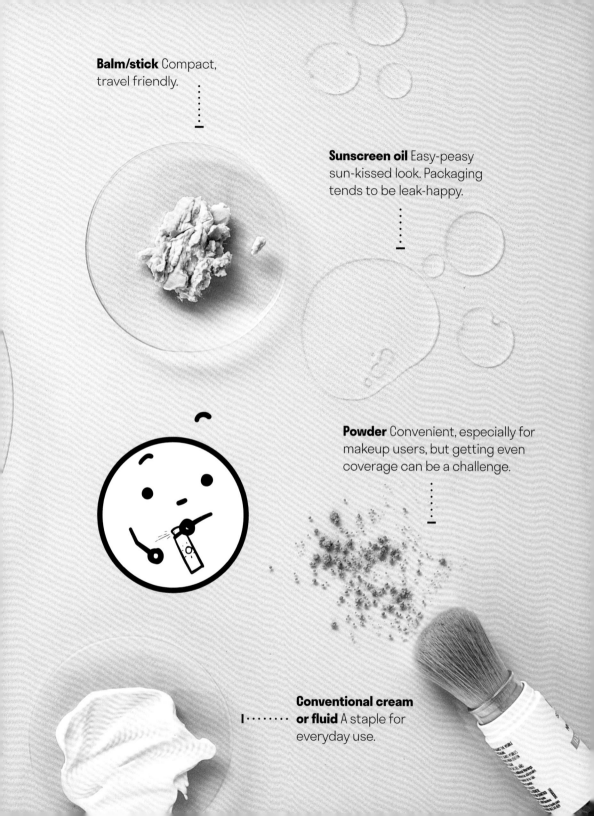

Balm/stick Compact, travel friendly.

Sunscreen oil Easy-peasy sun-kissed look. Packaging tends to be leak-happy.

Powder Convenient, especially for makeup users, but getting even coverage can be a challenge.

Conventional cream or fluid A staple for everyday use.

IN PRACTICE

Sunscreen is that finicky product that's a must-use but unfortunately has all sorts of texture and application issues. It can often be too greasy, shiny, smelly, or tacky, or make you look like a ghost. And, while you may want to try to aim for the highest SPF possible, we recommend finding sunscreens that you can tolerate and don't mind applying daily. A lower SPF that you apply daily will always be more powerful than the high SPF you only apply every once in a while. We recommend consulting our sunscreen product gallery on the previous page.

KEY THINGS TO KEEP IN MIND

The average human needs a shot-glass full of sunscreen for full-body protection.

SPF 30–50

Try to hit this SPF range.

PA++++

Broad Spectrum

UVA

Make sure you're getting UVA protection.

40 or 80 minutes

Planning to get some water time? Keep track of the time limit and reapply!

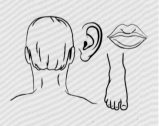

Common forgotten body parts that miss out on sunscreen protection.

Snorkeling? Instead of benzophenone-3 and methoxycinnamate, consider a mineral sunscreen.

To troubleshoot and narrow your sunscreen hunt, there are four general areas you can focus on:

1. **Chemical vs. mineral:** Sunscreen filters are divided into two categories: organic (chemical) and inorganic (mineral) filters. While chemical sunscreens are liquid filters that only absorb UV rays, mineral filters are powders that mainly absorb but can also reflect and disperse UV rays. Chemical sunscreens happen to have a better texture and less of a chance of leaving a white cast. But they aren't for everyone, as they can be a source of irritation for some. To make a decision here, it's important to get to know your ingredients. Good thing we've provided a table of sunscreen filters on pages 98–99.

2. **Skin type:** There's no doubt that oily/acne-prone skin types really struggle with the greasy sunscreens. We recommend looking to Asian and EU brands for lighter textures. The truth is, the US is quite limited in the approved sunscreen filters we can use, which means the sunscreens with the best textures are often sold outside of the country. In fact, we've often found that sunscreens with octinoxate tend to be lighter—just don't use them when you're swimming near the Great Barrier Reef. Otherwise, minerals can be hit or miss. If you're trying to go mineral, zinc oxide is the safer bet here. To avoid the white cast, use chemical sunscreens—otherwise, it may be time to go look for nano versions of these mineral suncreens.

3. **SPF value:** We typically recommend using anything from SPF 30 to SPF 50. We find that SPF 50 is a sweet spot between good protection and a lighter texture. But there are times when an SPF 30 is more ideal. If you're struggling with the white cast, texture, or sensitivity, we want to reemphasize that a sunscreen you'll apply more often will always be more powerful than a rarely used SPF 50 you hate. Remember, texture is king.

4. **To layer or not to layer?:** Sometimes layering can't be helped. If you plan to layer, do what you can to preserve the sunscreen film and ensure good protection. Make sure sunscreen is the last step in your routine, and allow enough drying time between your makeup foundation and sunscreen.

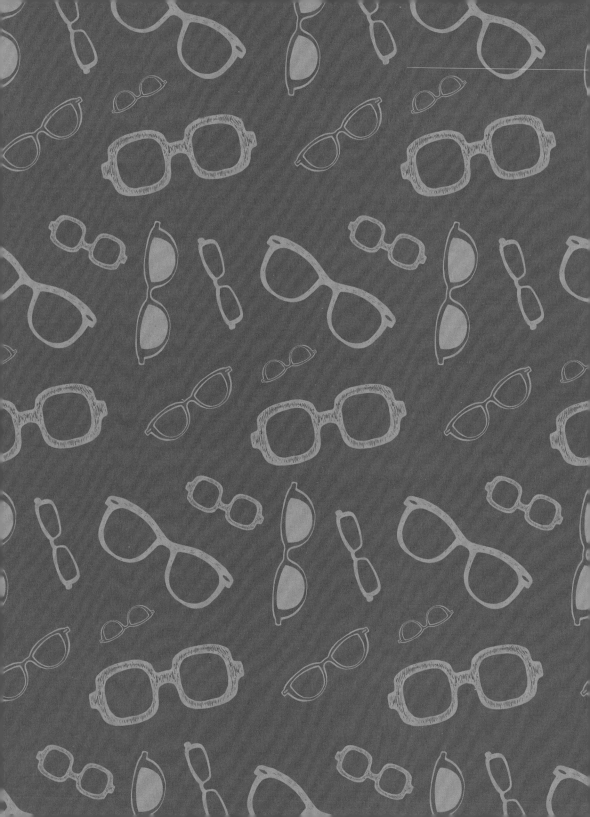

SUNSCREEN APPLICATION

Phew! We just wrapped a really long sunscreen chapter, but there's definitely more info to go through. There are a lot of scary, attention-grabbing headlines when it comes to using sunscreens. While most of them are a lot more hyperbolic than truthful, the real issue with sunscreens is that most people are not applying them correctly. So, now that we have the fundamentals of sunscreens covered, it's time to get into shopping for the one that works for you and learning the right way to apply these products to ensure you're getting the proper protection needed to prevent skin cancer and photodamage.

HOW ARE SPF VALUES TESTED?

Nowadays, you can find sunscreens in probably every formula type you can think of—lotions, creams, tinted, oils, sprays, mousse, sticks, and even powders. But is there an ideal type of sunscreen formula? Is this a situation where you shop based on your skin type or lifestyle?

To address these questions, we first have to look at how SPF values are tested. Every sunscreen—stick, powder, lotion, and so forth—must have its labeled SPF value tested by applying a set amount of product by weight over a set skin surface area. That's 2mg of product/cm² of skin, to be exact. Regardless of formula type, the *weight* of sunscreen you apply is absolutely crucial to getting that advertised SPF protection on the label. This is why one of our chemists' Golden Rule of Sunscreens is "Texture is King." The reality is, you're more likely to slather on enough of a sunscreen that feels great on your skin than a smelly, greasy, shiny goop of a sunscreen.

Different Sunscreen Formats are Deceiving!

Remember, sunscreen protection is measured by weight. But! Powders, oils, emulsions! All of these formula types have different densities. That means for products like oils and powders, you'll have to apply more than lotions and balms to get the same level of protection. Here is the same weight in five sunscreen formats.

| Gel | Stick | Powder | Oil | Cream |

In addition to texture, formula type also plays an important role. There are sunscreen formats that encourage you to apply close to the "correct amount." The density, spreadability, and how quickly the product sets can greatly impact how much sun protection you're *really* getting. See the picture below on the left.

Here, we measured out the same weight of a few drastically different formula types. That's right! Every beaker you see in this picture offers the same amount of protection to the same surface area of skin. This just goes to show how different your use experience might be, depending on the formula.

THE RIGHT WAY TO APPLY SPF

It's a well-known fact that a majority of people drastically underapply sunscreen. Studies have shown that most people apply just one-third to one-half of the required amount to get the claimed coverage. But, hang on, who actually knows their own skin surface area down to the cm^2 to calculate the proper SPF protection? Browsing on the interwebs, you can find some general rules and guidelines to help you roughly gauge how much sunscreen is needed. There are two popular guidelines you'll often come across. The first is the two-finger rule, where you apply a strip of sunscreen on two fingers to cover your face and neck. The second method is the simple ¼ teaspoon measurement, where ¼ teaspoon of sunscreen is sufficient for face and neck. But are these actually good rules of application?

Not to worry. We tested these out ourselves for you! We busted out our extra-precise laboratory scale, recruited our trusted friends and families, and put these rules to the test. We trialed every product type there is: sticks, powders, lotions, foam, oil. We weighed them out per application, and even put them through torture tests. By no means was this a rigorous study, but there are still a few solid takeaways to help ensure proper sunscreen application.

Why is sunscreen texture such a challenge, particularly in the US?

Due to regulations in the US, we're limited to a much smaller pool of approved sun filters to formulate sunscreen. The rest of the world has sadly moved on to more elegant textures thanks to improved sun filters like the Tinosorbs. Yes, this means the US is still stuck with filters from the Stone Age. The most problematic part of this limitation is that here in the US, we only can rely on one chemical filter for UVA protection: avobenzone. Even more tragic, avobenzone is a challenging ingredient to work with in terms of texture and stability. In fact, almost all of the filters US chemists are able to use don't feel great on the skin, and their stability problems are the stuff of chemist nightmares.

But wait! What about physical filters like zinc oxide, which can also offer broad-spectrum protection? In recent years, physical (aka mineral) sunscreens have enjoyed a major rise in popularity. That said, these are even messier in terms of texture. Mineral filters, such as zinc oxide and titanium dioxide, are the main culprits for the dreaded clown-worthy white cast. In our experience, the mineral sunscreen field is even more prone to unusable duds in terms of texture than chemical sunscreens.

All in all, sunscreens are just inherently difficult to formulate. The two of us can tell you that working on a "tear your hair out" sunscreen project is practically a rite of passage for many skincare chemists. These are incredibly finicky formulas. Changing even a harmless, vanilla ingredient can greatly impact the sunscreen's final UV protection. So if you have a holy-grail sunscreen that you love, you can thank a hardworking chemist who probably had to go through at least eighty-seven different variations for the one project to get it just right.

With sunscreens, "more is more!"

"A little bit goes a long way" is NOT a flattering trait for a sunscreen! Regardless of the product we were testing, when applying the proper amount, it always felt quite substantial on the skin. Even the best-performing formulas took a bit of time to rub in evenly. If your sunscreen takes less than five seconds to rub in, you're likely not applying enough!

Three fingers are more reliable than two fingers

This is really just an extension of "more is more." We found that the two-finger application method often yielded unreliable results. You simply have too many variables that would affect the amount. Everyone's finger length is different, the nozzle size changes how much sunscreen is dispensed, and how hard everyone squeezes the bottle is also quite different. All in all, if you fully glaze two fingers it *could* be a decent estimate for how much sunscreen you need for the face and neck area. However, simply using three fingers may be the better and safer estimate.

This only applies to classic sunscreen fluids, lotions, and creams. As you can probably imagine, with sprays, sticks, and powders, it just doesn't quite work.

A quarter teaspoon is a pretty good estimate!

Surprisingly, measuring out a quarter teaspoon of sunscreen is actually a pretty solid estimate for sunscreen use. There are some variables in terms of formula density and each individual's face and neck surface area, but it is much more consistent than using your fingers to estimate. Of course, it'd be a bit weird to take a measuring spoon with you everywhere you go. We would recommend measuring out a quarter teaspoon amount at home so you can visualize what that amount of sunscreen looks like in your palm as a future reference. This is a particularly useful way to estimate your sunscreen use with thinner, more fluid formulas that would run right off your fingers. This method also doesn't apply to sprays, powders, sticks, and mousse formulas.

BATTLE OF THE SUNSCREEN FORMULAS

Other than traditional more creamy sunscreen emulsions, there are a lot of options from which to choose. It might feel like these different formats are better fits for your lifestyle . . . but are you really getting the coverage you think you're getting?

Tinted Sunscreen or Makeup with SPF

When done well, these types of products can be absolute lifesavers. They are a great way for you to get some basic daily SPF protection without compromising your makeup. But the right texture to target here might not be what you typically look for in a makeup product. Foundations often come with the "little bit goes a long way" strategy, which is the opposite of what you need in a sunscreen. That means textures that have a lot of glide and coverage may prevent you from applying enough product to get the proper SPF protection. Textures with lighter yet buildable coverage are the way to go here.

Sunscreen Powder

Sunscreen powders at first glance may seem very compelling. They seem travel-friendly, layer-friendly, makeup-friendly, oily-skin-friendly, and, more importantly, reapplication-friendly. Sadly, we have to give this format a big, resounding thumbs-down. Remember how all sunscreen SPF is measured the same way? This applies to powder formats, too. This is a *lot* of powder—much more than anyone would reasonably want to apply on their face. We're talking about a walking powder-puff level of powder. Not to mention powder-format sunscreen is missing a lot of the ingredients in conventional sunscreens to enhance performance, such as oils and film formers that help evenly disperse sun filters. Ultimately, this might be okay to use for some extra protection on days when you're getting minimal sun exposure, but we absolutely would not recommend solely relying on a sunscreen powder if you're heading to the beach.

Sunscreen Stick

Sunscreen sticks are another format that seems very convenient when you're on the go. They can be a great way to touch up on your sunscreen, but, sadly, we have found that one swipe of sunscreen balm doesn't deposit anywhere near the targeted $2mg/cm^2$ quantity. If you use a sunscreen balm to touch up, try to do at least two passes. We also recommend rubbing in the sunscreen post-application, just to make sure you didn't miss a spot.

Spray

Many people enjoy the convenience of using spray sunscreen for full-body coverage. It can be a very convenient, reliable, and refreshing sunscreen format. But after you spray it on, don't forget to fully rub the sunscreen into your skin. It still needs to have a nice, even film for it to give proper protection. The most recommended method used in SPF testing is to spray the sunscreen into your hand first, then rub on your skin. Even if you prefer to spray directly onto your face, don't forget to rub it in!

Mousse

Recently, we've noticed a few more sunscreen options that come in mousse form. Surprisingly, these have fared pretty well! A good dollop dispenses a sufficient weight of sunscreen to give you the right level of sun protection. The mousse texture also helps the application experience feel less heavy, although the finish is still on the nourishing side. This may be a great option for those with dry skin. Aim for a golf-ball-sized amount for face and neck.

Hybrid moisturizers

The best format for lazy folks who don't want too many layers. Be sure you apply enough of the product—it's important to remember you're applying based on the required sunscreen amount and not the moisturizer amount.

BONUS! PRO TIPS ON REAPPLICATION AND WASH-OFF

Reapplication is probably the worst part of sunscreen, since it's usually at a pretty inconvenient time, when you're sweaty, sandy, or both! The slightly cumbersome method for a cleaner reapplication is to have a makeup wipe handy. We felt this was incredibly necessary for tinted sunscreens because applying a second coat of a heavily pigmented formula was not fun. Not only do you look pretty silly, but we're talking about dealing with some pretty gross textures and ruined laundry as well.

As far as washing off sunscreen goes, we've tested many with our gentle Blank Slate Gel Cleanser and most did pretty well, except for the tinted formulas. However, when you're outside all day and reapplying consistently, there can be some gnarly buildup by the end of the day. Think of all the sweat,

Rubbish Alert!

A serious word of warning: Avoid concocting homemade sunscreens.

Sunscreens are considered to be OTC products under FDA regulation—what this means is that anything you find on store shelves has been tested for efficacy. A homemade sunscreen recipe often uses zinc oxide, an effective broad-spectrum UV filter. However, without a proper formula and the right manufacturing equipment, the DIY brew might not disperse the filters properly in an even film, leaving you with patchy, lackluster sun protection. We have seen plenty of "safe," "chemical-free" sunscreens sold at hipster markets that are questionably preserved and already starting to separate into multiple layers. Skin cancer is no joke! Let's leave the sunscreen formulation to the professionals.

grime, and sunscreen layers you've accumulated! Double-cleansing with an oil cleanser before your standard cream or gel cleaner is still the most reliable way to ensure that your skin is thoroughly clean by the end of the day.

Decoding the Fitzpatrick Scale

Have you ever wondered how dermatologists classify skin tone? No? Well, we'll tell you anyway.

The Fitzpatrick scale classifies skin color into six categories. This may seem almost absurdly general (thank goodness Dr. Fitzpatrick didn't try to come out with a line of foundation), but what it actually represents is how your skin reacts to sun exposure. Fair-skin types, on the low end of the Fitzpatrick scale, burn easily and have a higher cancer risk. Darker-skin types tan more easily instead of getting sunburned. So, as you can imagine, good sunscreen studies include a blend of skin tones on the scale.

It's important to note, though, that just because a darker skin tone has less chance of burning, that doesn't stop UV rays from damaging the skin. Stay on that sunscreen habit to keep skin cancer and photoaging away!

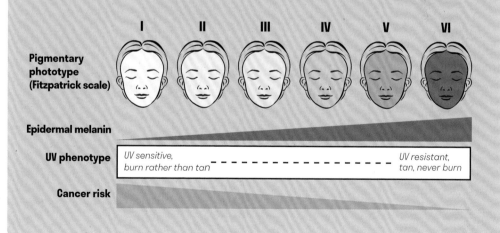

| | I | II | III | IV | V | VI |

Pigmentary phototype (Fitzpatrick scale)

Epidermal melanin

UV phenotype — *UV sensitive, burn rather than tan* — — — — — — — — — — — — *UV resistant, tan, never burn*

Cancer risk

LEVELING UP YOUR SUN PROTECTION

L ove the great outdoors? Well, slathering sunscreen over your entire body isn't the only option you have! You can further protect yourself with physical gear.

UV-rated clothing: Fun fact! Sun-protective clothing gets tested for sun protection, too, and gets a UPF rating. This is a number that indicates both UVA and UVB protection. The UPF rating means that, of a given amount of UV, only one part gets through. For example, UPF 10 allows one part in 10, UPF 20 would let through one in 20, and so on; the higher the UPF number, the better the protection. This is important for extensive outdoor time!

Sunbrellas: We often hear questions like, "If I sit under an umbrella and don't apply sunscreen, I should be protected, right?" Funny enough, someone actually compared sunscreens to sunbrellas in a clinical test. Johnson & Johnson did a clinical study in sunny Texas that looked at just that. They had people either apply sunscreen or sit underneath an umbrella. They found that those who sat underneath the umbrella experienced more tanning and sunburns than those who just used sunscreen. Why? Even when you sit in the shade, your body is still being exposed to UV, since UV is reflected off surfaces like sand and water. So, to sum up: Being in the shade of a beach umbrella may help, but it's not fully sufficient sun protection.

Sunscreen Can of Worms

One controversy with chemical sunscreens:

On the whole, chemical sunscreens aren't the "toxic chemicals" that you may see getting criticized in clean-beauty circles. As noted earlier, they're pretty simple carbon-based sun filters, hence, their alter ego as "organic." They generally work by absorbing UV and dispelling that energy as heat. Chemical sunscreens tend to have better textures and be more friendly to oily, acne-prone skin types. There's also a much lower chance of chemical sunscreens leaving a white cast on darker skin tones. But they aren't perfect. These have a slightly higher potential for irritation, and a couple of them draw concerns about impact on coral reefs.

In addition, as we write this update in 2024, there have been recent findings regarding chemical-sunscreen absorption in the blood. The FDA tested seven chemical-sunscreen filters and found that all of them were absorbed into the bloodstream well over the declared safety limit of 0.5ng/mL (half a nanogram per milliliter).

As alarming as this may sound, there's no real conclusion here except that a higher concentration is in the blood. We don't know whether this is actually unsafe. The 0.5ng/mL limit was more of a general safety guideline, but this is now in question. Does that mean each ingredient will require a specific limit?

Despite these learnings, the sun is still the real enemy. The FDA is not saying you need to abandon chemical sunscreens—in fact, we should see this as the agency truly investigating the safety of sunscreen products. We will be eagerly awaiting more updates. Meanwhile, keep to the strategy: Use sunscreen!

MEET THE SUNSCREEN INGREDIENTS

Paying attention to sunscreen ingredients can help you eliminate potential troublemakers that your skin isn't happy with and help you discover your sunscreen-filter loves. To help, we've provided a quick ingredient guide below to help you familiarize yourself with common sunscreen ingredients around the world. Use this guide to get acquainted with your options and to inspire your next sunscreen purchase.

	US-Approved Organic (Chemical) Sunscreen					
Trade name	Avobenzone	Homosalate	Octisalate	Octocrylene	Oxybenzone	Octinoxate
INCI name	Butyl Methoxydiben-zoylmethane	Homosalate	Ethylhexyl Salicylate	Octocrylene	Bezophe-none-3	Ethylhexyl Methoxy-cinnamate
UVA/UVB?	UVA	UVB	UVB	UVB	UVB	UVB
Conclusion	From this chart of US chemical filters, you'll see that we're basically stuck with Avobenzone for broad-spectrum protection. Moreover, all the US chemical filters are oil-based ingredients. What this means for us poor chemists is that it's an uphill battle to create light, pleasant, layerable formulas with a bucket of thick, finicky, sometimes unstable oils. Yuck.					

* International Nomenclature of Cosmetic Ingredients.

Asia/EU-Approved Organic (Chemical) Sunscreen

Trade name	Tinosorb S	Tinosorb A2B	Uvinul T 150	Uvinul A Plus	Ecamsule, Mexoryl SX
INCI name	Bis-ethylhexy-loxyphenol Methoxyphenyl Triazine	Tris-Biphenyl Triazine	Ethylhexyl Triazone	Diethylamino hydroxybenzoyl hexyl benzoate	Terephthalylidene dicamphor sulfonic acid
UVA/UVB?	UVA/UVB	UVA/UVB	UVB	UVA	UVA
Conclusion	Ahh, a list of all the good things we can't have in the US. Some of these ingredients can even be dispersed in water. This is why you can find lighter sunscreens in Europe and Asia, rather than the greasy American ones. The full list of approved ingredients is even longer; we just listed some of the most common ones you'll find in European and Asian sunscreens.				L'Oreal-patented ingredient you can find even in US sunscreens.

Inorganic (Physical or Mineral) Sunscreen

INCI name	Titanium dioxide	Zinc oxide
UVA/UVB?	UVA/UVB	UVA/UVB
Conclusion	The two physical filters are approved for use worldwide and are often recommended for those with sensitive skin. On top of that, they are free of controversial claims like environmental harm and absorption into the bloodstream. The downside? These tend to leave a clownish white cast. The "nano" grade of these powders can help minimize this white-cast problem. There are people concerned about the health effects of ultra-small particle sizes, but those claims have not been substantiated by science.	

CHEAT SHEET
Sunscreen Summarized

Chemist Guidelines

- The averge person doesn't apply enough sunscreen. Apply MOAR.

- Follow the instructions and apply every two hours. If you're spending a lot of time in water, apply based on the water-resistance limit.

- Aim for the SPF 30–50 range.

- Make sure you're getting both UVB and UVA protection—look for "broad spectrum," "UVA," and "PA" labels.

Recommended Starting Point by Skin Type

Oily Chemical sunscreens with a thinner texture.

Dry Thicker chemical sunscreens without high levels of alcohol or powders. Avoid mattifying sunscreens.

Sensitive Look for a fragrance-free mineral sunscreen.

Chemist Hacks and Pro Tips

- White-cast issues? Try a chemical sunscreen, or if you have sensitive skin, you can use an SPF 30 mineral sunscreen.

- Are you layering your sunscreen and it's causing pilling issues? Try to find a sunscreen with fewer silicones.

PERSONAL TAKE

By now, you can see that there's no one standard answer to skincare. It's all about your current skin situation and what works for you. Here are some of our personal notes on this category.

Gloria

I've had pretty garbage sunscreen habits all the way until . . . well, basically, until I learned the importance of actually using sunscreen after becoming a cosmetic chemist. Given that my skin is dry, I'm not as picky about greasy textures so long as they dont "shine." My sunscreen Achilles' heel, though, is mineral formulas. Mineral filters just tend to make my dry skin feel even dryer, more powdery, and did I mention, dry? With all the recent drama around chemical filters, I am curious about what new, improved sunscreen formulas are on the horizon!

I have struggled with sunscreens my entire life. I don't need to look any more greasy, and yet some of the dry-touch ones seem to irritate my acne even more. It would make sense for me to go mineral, but a lot of those make me look ghostly, and why do they have to be so tacky?! Grr. I've found the Asian sunscreens work best for me for texture, finish, and white cast. But I will also mention that some Asian sunscreens "tone up" (whiten) skin, so it still takes some trial and error.

Victoria

SUNSCREEN FAQs

 Q: The sun scares me. Do I need to use SPF 100?

A: Generally, if you're looking for extra protection, SPF50 or 50+ is all you need. At that point, you're already able to block more than 97% of the sun's rays. Anything more means you'll have to come to terms with a much heavier, greasier texture that often leaves a white cast. We're all about compliance here. Wearing a sunscreen that you like daily will always be more powerful than wearing an SPF 100 whenever you can tolerate it.

 Q: Do I need to reapply sunscreen at the end of my workday even though I was only in the sun for my ten-minute morning commute?

A: Sunscreen doesn't work like batteries, where you have a reserve of sun protection. The sunscreen film is meant to hold up only for two hours, and that's it. Definitely reapply sunscreen for your after-work commute.

 Q: How should I layer my sunscreen?

A: Make this the last step of your routine. Applying anything over it really runs the risk of interfering with the sunscreen film and thus diminishing sun protection.

4

Q: How do I reapply sunscreen with makeup?

A: Great question that comes with a not-so-great answer. Sunscreen powders have been formulated because of this problem, but we don't actually recommend using them. It is very difficult to quantify the proper amount of powder sunscreen. Instead, we would consider a sunscreen face spray over powders. But the optimal solution would be to use a CC or BB cream instead of foundation.

5

Q: I'm worried sunscreens are the source of my breakouts!

A: Seek out better textures—the European and Asian formulas are lighter!

6

Q: I only used my sunscreen once, but it's expired. Can I still use it? It feels like such a waste to throw it out.

A: Don't do it! Sunscreen formulas are finicky enough to formulate already—when it's time for it to go, it's gotta go.

7

Q: My sunscreen seems to be leaking oil.

A: Once again, time to dispose of the sunscreen, just as in question 6.

Section 2: TREATMENTS

BEYOND THE BASICS

We classify "treatments" as all of those ingredients and products that are meant to bring long-term benefits to your skin. Think of your cleansing, moisturizing, and sun-protection routine as daily, fundamental upkeep. On the flipside, treatments are products that target core skin concerns such as wrinkles, pigmentation, sagging, tone, and texture. This category is doused in glamour (*cough* money). It's also one of the most confusing categories of skincare. At the end of the day, every fancy-shmancy serum sounds like the holy grail, and most don't even come close. The reality is that actives are not just fancy, a lot of them are also fussy. It takes know-how to formulate a good treatment that will hold up in shelf life, but you also need a little bit of knowledge to use these products effectively. Good thing you have us, huh? Let's go beyond the basics!

THE MESSY WORLD OF ACTIVE INGREDIENTS

It turns out that behind every treatment or serum that targets those pesky skin concerns are hero ingredients we chemists call actives. Active ingredients are molecules that deliver long-term benefits to the skin, allowing brands to make claims of reclaiming youth and glow. The skincare industry is just *filled* with these powerhouse ingredients, but, unfortunately, it's difficult to make heads or tails of what these guys actually do for your skin. Since many active ingredients' benefits are not well classified, you'll find that some have multiple benefits, while others seem to be grasping at straws as to what they actually do. In fact, our little guide below gives you an idea of just how confusing active ingredients can get!

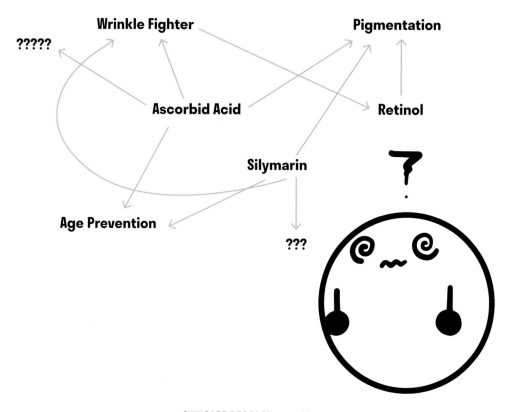

Even just trying to make sense of a few actives can make you go a little cross-eyed. It doesn't help that there are thousands of active ingredients out there—some good and some absolute garbage. How did it get this way?

It turns out there are a few ways active ingredients can make their way into your favorite serum:

1. **The exotic heritage—story:** Sometimes a flower extract just has that sexy origin story to convince you that it's the miracle worker that's been missing from your life all along. So, let's throw it at acne, inflammation, and hyperpigmentation to see how it does. Claims often make products sound like they're the holy grail, but actual performance varies widely!

2. **The trickle-down path:** If an ingredient has a positive reputation in food, medicine, or aerospace engineering (not a joke!), why not try to put it in a skincare cream? Actual skin benefits are often overstated.

3. **The shotgun approach:** There are also molecules designed for skin benefits, and they get tested for *every* skincare benefit you can possibly think of. Hence, this is why an ingredient can have many overlapping benefits, adding to the chaos.

As you can see, there's already a lot to navigate when trying to choose the right active ingredient without even considering your specific skin concerns and skin type. This is further complicated by the fact that the active-ingredient world is teeming with products that are ineffective at best and snake oil at worst. We could go through every claimed skincare active out there, but this book would quickly morph into an encyclopedia, and we'd like to save a few trees.

So, worry not! Instead, let's simplify the overall landscape and look at what are *truly* worthy treatments, with proven efficacy. We'll be answering the following three questions:

1. **What is considered proven efficacy for an active?**
2. **What, in the chemists' opinion, are the fairest actives of them all?**
3. **How do you incorporate these key active treatments into your routine?**

IN SEARCH OF REAL EFFICACY

So, what do we mean by "proven efficacy"? There are actually a ton of different ways scientists can test active ingredients to verify how and what they really do for your skin. Here are the most common tests, and what they mean:

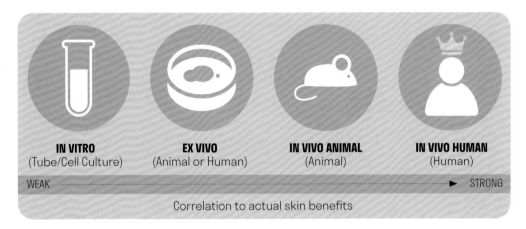

IN VITRO	EX VIVO	IN VIVO ANIMAL	IN VIVO HUMAN
(Tube/Cell Culture)	(Animal or Human)	(Animal)	(Human)

WEAK ·· ► STRONG

Correlation to actual skin benefits

In a petri dish (in vitro/in tubo): These active ingredients are being tested in a test tube or a cell-culture petri dish. Sometimes this method is used to figure out how an ingredient interacts with certain skin cells, such as fibroblasts. This testing style is also used to see if an ingredient can act as an antioxidant to defend against free radicals. The downside is that it in no way mimics how these ingredients are absorbed through the skin to reach the targeted cells. Think of this as an initial litmus test, a green light/red light on whether this ingredient should proceed to more testing.

On removed skin (ex vivo): Active ingredients, alone or in formula, are tested on a patch of removed skin, typically from pigs or humans. Pig skin is the animal skin of choice, since it happens to be very similar to our skin. Human skin is also sometimes used, and can be sourced from plastic surgery "leftovers." The upside of this method is that this is a much better way to understand how an ingredient or formula interacts with and absorbs into skin, since we're using, well, actual skin. The big downside is that you can only keep skin "fresh" for so long, so these tests are usually limited to two weeks max, which isn't a lot of time to really understand the long-term benefits of actives.

In vivo (animal): Ingredients are tested on live animals. Whoa! Before you start grabbing your pitchforks, most in vivo animal testing nowadays is kept to a minimum, and usually only used in academic settings. It can help us understand how an ingredient may protect against UV light or accelerate wound healing, but, ultimately, this still isn't great, because translating animal data to humans isn't straightforward. You can rest easy, as nowadays there is less and less animal testing done in the beauty industry. We're animal lovers too!

In vivo (human): We've made it to the most ideal testing scenario—testing actives on actual human subjects! But, unfortunately, this is also the most expensive scenario, which means not a lot of active ingredients actually make it through to a good in vivo human clinical test. Because of how expensive this type of testing is, the data quality can also vary greatly. At the end of the day, testing skincare ingredients isn't nearly as rigorous as testing medicine. So, as we evaluate ingredients, we often feel like beggars, scrounging around for decent data to help us assess their efficacy! Here are two important criteria we consider in evaluating clinical tests:

1. **Number of test subjects:** We typically look for at least 30 test subjects. It's actually not a lot, but we consider this decent compared to some studies, which may include a whopping five subjects.

2. **Whether a test is placebo-controlled:** This means studies compare skin treated with the active ingredient to skin treated with a blank formula. This is *the* most direct way of figuring out what an active ingredient does for your skin. Without a placebo, you have no baseline of comparison, so studies that do this are near and dear to our hearts!

Phew! As you can see, there's a lot that goes into validating an ingredient. And, sadly, very few active ingredients in the skincare market are backed by solid, chemist-drool-worthy clinical studies. Which brings us to . . . the skincare Big Four.

MEET THE SKINCARE BIG FOUR

Setting all the marketing smoke and mirrors aside, at the core of active ingredients are the Big Four. These ingredients end up covering a comprehensive variety of skin concerns, including wrinkles, pigmentation, texture, and even acne. So, let's meet your Big Four: the skincare ABCs, the bread and butter, the four knights . . . the four musketeers! (Okay, we'll stop.) These are ingredients that have been tested and proven over and over again in countless (quality!) clinical studies over the years. They are:

1. **Chemical exfoliants:** Acids such as glycolic, lactic, and mandelic, which gently dislodge dead skin cells that have overstayed their welcome.

2. **Retinoids:** The OG do-everything category of ingredients.

3. **Vitamin C:** Staple antioxidant for long-term aging prevention.

4. **Niacinamide:** Best supporting ingredient for brighter, more hydrated skin.

Each of these comes with a long history of proven benefits for many skin concerns but can be quirky and finicky to formulate with. Choosing the right products and effectively integrating them into your skincare routine can be tricky. In the coming sections, we'll explain how they interact with skin, what benefits they bring, how these should be used, the length of time needed to see benefits, and general tips on how to incorporate these ingredients into your routine.

Beyond the Big Four

"You're only covering four actives?! Is that . . . all?"
You're right—there's so much more out there. (Peptides! Green tea! Stem cells!) And even though understanding and evaluating these other actives can be such a clusterfork, part of the fun of skincare is discovering and trying out new products and ingredients. To wrap up this section, we'll run through some of the other more popular actives out there and help decipher what actually works, and what's thrice-distilled, 90-proof BS.

The Chemists' Guide to Treatments

Before we jump into the meat of the treatments section, here are a few Chemists' guidelines to keep in mind:

1 **Concentration:** If there's anything that you should take away from this book, it's that if you want things to work, you need the right amount. The dose makes the elixir! This is especially relevant with heavy-duty skincare ingredients designed to tackle all your skin concerns.

2 **Patience:** Good things take time. These active ingredients will, on average, need four weeks or so of consistent use to see any sort of improvement. (Some even need a couple of months!) As tempting as an overnight miracle sounds, that would really be too good to be true.

3 **Irritation:** Skincare makes it easy to have a candy-shop mentality! Antiwrinkles, yes! Glowing skin, yes! And suddenly you've stacked ten actives in the same routine and you've hit Irritation Town. That is not a place you want to be. You'll end up spending your entire routine trying to get your skin back to normal, instead of focusing on the good stuff.

4 **Layering:** Generally speaking, you want to layer from water-based products to oily products. The treatments section is also where you should think about "Less is more." A solid four- or five-step routine should cover all your bases, especially if you're using multitalented anti-wrinkle ingredients that can target more than one issue. Less is more!

5 **Molecule Matters:** Let's use an example: vitamin C has all sorts of derivatives, but you'll see we spend a good chunk of text explaining how ascorbic acid is the champion vitamin C out of all of them. So, the type of vitamin C does matter, and not all vitamin Cs are created equal.

CHEMICAL EXFOLIANTS

We've all been told about the importance of exfoliation since the days when St. Ives Apricot Scrub was all the rage (a truly age-revealing reference). The perception has always been that you can scrub your way to glowing, baby-soft skin. But physical exfoliation isn't the only (and sometimes not really the best) way to effectively slough off dead cells and keep skin turnover healthy. Meet the chemical exfoliants: ingredients that weaken the chemical bonds holding those stubborn, overly attached dead cells together and return skin to its smooth, glowing self. Here, we'll cover the entire exfoliation landscape, and how to rein in chemical exfoliants for your skin's "Aha!" moment.

WHY EXFOLIATE?

For skin to stay healthy and look youthful, it has to shed constantly. But as we get older, this cell-turnover process can start slowing down, leading to dull, rough skin. This is where chemical exfoliants come in, helping to kick out those old cells that have overstayed their welcome.

Healthy Skin Has Healthy Desquamation

Desquama-who? Desquamation is the process of fresh skin cells moving up and old cells shedding. This is essentially your skin's turnover rate. In healthy stratum corneum, the cell-turnover cycle is about two to three weeks. Your entire epidermis turns over in about seven weeks!

YOUR SKIN CONSTANTLY SHEDS!

* Corneocytes don't actually use parachutes.

Skin's turnover rate is crucial to its health. Desquamation can slow down due to aging, environmental stressors, and dehydration. When this turnover process slows down, it starts a crappy, vicious cycle of rough texture ⇢ compromised barrier function ⇢ dehydration ⇢ further cell-turnover slowdown. Meaning, your skin becomes rough, scaly, and dull. Eek!

Enter Chemical Exfoliants

This is where chemical exfoliants come in! *Chemical* means they are acids that weaken the bonds (proteins called corneodesmosomes, should this ever come up at trivia night) between dead cells that have overstayed their welcome, which helps your skin return to a healthy turnover rate. Shedding the top layer of cells has the short-term benefit of smoothing skin texture. In the long run, some of these ingredients can help tackle pigmentation, acne, and even—*gasp*—collagen production! Intriguing, yes? So, let's figure out which one is right for you.

A Very Brief History of Chemical Exfoliants

Have you ever wondered who was the brave soul who first put acids on their face? Turns out, there's a long, long history of people using such substances for skin beautification. In ancient Egypt, women would use sour milk, rich in lactic acid, to achieve that glow. Greeks, Romans, and various nomadic cultures all have had their own unique formulas for skin exfoliation. During medieval times, vinegar and wine were esteemed as beauty treatments. We're doubtful that they were used with the intention of exfoliation, but we can consider this one of the first mentions of a true chemical exfoliant, since wines typically sit in the pH range of around 2.5–4.5 and contain tartaric acid, which is . . . an alpha-hydroxy acid! (Albeit not a very good one!) Good thing we now have better alpha hydroxy acids, so save the wine for drinking!

But chemical exfoliations didn't start getting serious until the late 1800s, when ingredients like phenols, resorcinol, and salicylic acid started becoming popular. The first chemical peels targeted skin conditions like melasma and unwanted freckles. Surprisingly, AHAs were a little bit of a late bloomer, and didn't become popular until the 1980s.

MEET YOUR CHEMICAL EXFOLIANTS

There are three types of chemical exfoliants, all mild acids. They're divided by their molecular structures, but we'll save you from this chemistry snoozefest. The good thing is that these categories can also be characterized by their function.

Chemical exfoliants will help slough off those stubborn, past-due skin cells and kick your cell turnover into gear. One cool thing about these ingredients is that they can bring the almost immediate benefit of smoother skin texture. And that's not to mention all the long-term benefits, such as helping with pigmentation and acne! So, let's take a look at the options.

The Bread and Butter: Alpha Hydroxy Acids (AHAs)

The AHA category includes glycolic, lactic, mandelic, malic, and tartaric acids. These are water-soluble, weak acids. Of these acids, we'll only focus on three: glycolic, lactic, and mandelic. The main difference between these three is their molecular size, which can help you choose which one to be your AHA soulmate. These have quite a long history of proven skin benefits for pigmentation, collagen production, and fine lines and wrinkles.

The Oily-Skin Specialist: Beta Hydroxy Acid (BHA)

Really, the only chemical exfoliant in this category is salicylic acid, which has the unique characteristic of being slightly oil-soluble. This unique trait allows salicylic acid to exfoliate down to the pore level. Add in the fact that it's also an antimicrobial and anti-inflammatory ingredient, and you've got an ideal candidate for inflamed, acne-prone, oily skin.

The New Kid: Gentle Polyhydroxy Acid (PHA)

Who dat? Yep, polyhydroxy acids are the new kid on the block. PHA typically refers to two molecules: lactobionic acid and gluconolactone, with gluconolactone being the much more common one in skincare. It's said to be an ultra-gentle e chemical exfoliation experience, and even acts as a humectant, keeping hold of water and hydrating skin. In fact, it's so gentle that it doesn't even make your skin sun-sensitive, unlike the others. (Okay— seriously, though, let's still play it safe and wear that sunscreen!)

There's even data on gluconolactone's potential to treat acne, and there will be more data to come. All in all, this is an ideal candidate for those who struggle with dry skin, sensitive skin, or just haven't had a lot of luck with the other exfoliants we've mentioned.

Rubbish Alert!

When you experiment with higher-level acids, light stinging is normal, but it shouldn't feel like you're trying to purify demons through your pores.

Something absolutely terrifying that we see all the time in the mystical world of Instagram is people taking chemical exfoliation way too far. Sometimes we see influencers buff then scrub then peel their faces until their skin looks more raw than Grade-A steak. Which, *owww*. When using acids at home, light, temporary tingling is perfectly normal. However, you should not be lobster-red and in pain afterward.

And what about those DIY home recipes? Specifically, what about apple cider vinegar and lemon juice? Good question! Apple cider vinegar has about 5% acetic acid, while lemon juice will have about 5% citric acid. It turns out neither of these acids has a lot of evidence bringing any topical benefits. Plus, pure lemon juice or vinegar can be too acidic for skin. So, let's stick to the store-bought products, shall we?

CHEMICAL EXFOLIANTS IN PRACTICE

Anyone can benefit from having chemical exfoliants in their routine. But how do you go about choosing a product that's right for your skin? We recommend the following steps: (1) Find your chemical exfoliant BFF molecule, (2) choose the right product type for you, and (3) troubleshoot your routine. Here's how.

1. Where Do I Begin?

When it comes to choosing your acids, size matters. The smaller the molecule, the more aggressive, yet effective it is. So, in the AHA landscape, in terms of efficacy, lactic acid is larger than glycolic acid, which is larger than mandelic acid. But if you've never used chemical exfoliants before, jumping right into that 30% glycolic peel is like trying to unicycle before you've taken the training wheels off your bicycle. Refer to the decision tree on pages 126–127 for help finding your ideal starting point and how to level up.

2. Which Product Is Right For Me?

Chemical exfoliants are super popular, and you can find them in all sorts of products. They are most commonly found in cleansers, toners, serums, creams, wipes, and masks. Wait—that covers just about every product type out there! Right off the bat, we wouldn't recommend incorporating acids through your cleansers. Just not the most effective way of getting that chemical exfoliating power. For all the other product types, there are three things you need to consider:

Concentration: Ah, yes! Like a broken record, we've hit our favorite Chemists' Commandment again: Make sure you're hitting those percentages. And for AHAs, these concentrations are pretty high. In fact, concentration is so key here that if you find products that don't disclose the concentration of acids, just walk away. It ain't worth the time.

> **For effective daily upkeep, here's what to look for on labels:**
>
> 5–10% glycolic acid
>
> > 8% lactic and mandelic acids
>
> 0.2–2% salicylic acid
>
> 10% gluconolactone

The product's pH: For AHA/PHA products, the lower the pH, the better it works. But of course, it shouldn't just be as low as can be. There's a balance between "good efficacy" and "Holy hell, it *burnnns*!" For most people, there's little chance of irritation if the pH is around 3.5, and you still reap the exfoliation benefits.

Support ingredients: Of course, every great formula needs a great supporting cast to really shine! Here are two ingredients you'll want to look for:

pH adjusters: These ingredients—which include sodium hydroxide, potassium hydroxide, and triethanolamine, among others—are crucial. Without them, your AHA pH can be *really* low. We're talking a pH of 1 or even lower. So, if a product doesn't use any adjuster, that's pretty sketchy. Our advice? Don't walk, run! Leave those products for dermatologists and aestheticians to use.

Soothing ingredients: Some products include "soothers" to mitigate skin reactions and make your AHA experience more enjoyable. If you want to add some soothing assistance, look for ingredients such as bisabolol (German chamomile extract), calendula, and *Centella asiatica*.

3. How Do I Incorporate AHAs into My Routine?

Now you're ready for your skin's AHA moment! (Teehee.) Here are just a few Chemists' guidelines for your exfoliation journey:

Apply in the right order: Chemical exfoliant acids are usually in heavily water-based formulas. That means that your AHA wipe, toner, or serum (pick one!) should fall into either step 1 or 2 of your nighttime routine, right after cleansing. Products with a concentration of under 10% AHA can be used daily. It you're sensitive to even a lower concentration, use it on alternate nights.

Consider a high-concentration mask: A great way to elevate your exfoliant game is to add a high-level (20% or more) product to your skincare arsenal. These higher-concentration AHA products should be used as rinse-off masks no more than once a week.

Support Products & Troubleshooting

Fun fact! Certain products you may already have can enhance your chemical exfoliation routine. The big three are listed below:

Clay masks and alcohol-based toners: Since these products remove excess oil from your face, they can be ideal before your acid product. Excess oil and grime can prevent these acids from penetrating, so by using a clay mask or toner beforehand, you can boost your AHA efficacy without having to reach for a higher concentration. Just don't leave the clay mask on for too long; a good five to ten minutes is all you need.

Petrolatum and balms: If you have a dry patch, a small cut, or just a slightly sensitive area, you can use petrolatum or a balm on those trouble spots before applying your treatment. That way, the acid won't reach those sensitive patches.

Pro Tip: As we mentioned, you can find exfoliants in all sorts of product types, which can sometimes lead to accidental irritation. Layering a salicylic acid toner, and a glycolic acid serum, and then wrapping up with a lactic acid moisturizer is just setting yourself up for overexfoliation. So, if you have a mystery case of skin irritation, double-check all the ingredient lists to make sure you're not accidentally overdoing the chemical exfoliants.

Physical exfoliants: Chemical exfoliants loosen up dead skin for that smooth, smooth glow. But as you age, your skin can get stubborn and refuse to let go of those old corneocytes, even with chemical exfoliation. Use a gentle physical exfoliant to mechanically buff those loose skin cells off. Something as gentle as a soft, bristled cleansing brush or konjac sponge works great. Say no to 50-grit, sandpaper-worthy, apricot-seed or walnut-seed scrubs. It doesn't have to be that way anymore!

Sunscreen: Last, but not least . . . remember sunscreen! Most of these acids can make your skin more susceptible to sun damage. Of course, whether or not you're using chemical exfoliants, you should always practice good sunscreen habits anyway. Sunscreen every day keeps the Botox away . . . ?

Other Exfoliants

"What's up with enzyme peels? And whatever happened to gel peels?"

Enzyme peels typically use bromelain, an enzyme found in pineapples, or papain, from papayas, instead of AHAs, BHAs, or PHAs. Enthusiasts say they get all the efficacy of the chemical products with none of the irritation. There's a study out there that claims 1% papain is superior to 5% lactic acid. We even found one claiming that roe extract (yes, you heard that correctly—fish eggs) is better tolerated and more effective than 4% glycolic acid. These studies are few and far between, so most of the "data" behind enzyme peels comes from manufacturers and product marketing. But, if your skin enjoys a glycolic peel, there's no need to ditch it for salmon-egg essence.

   Gommage, or gel peels were all the rage in the early 2000s. Remember those overly enthusiastic mall-kiosk workers rubbing a miracle gel onto your skin to produce gross little balls of "dead skin cells"? While some gommage peels do have low levels of AHAs, they are actually primarily physical exfoliants. The little balls are thickening fibers (usually cellulose) that help roll off loose dead skin in the most gentle way. As far as physical exfoliants go, this is as gentle as it gets. (Read: So gentle that they're not super effective.)

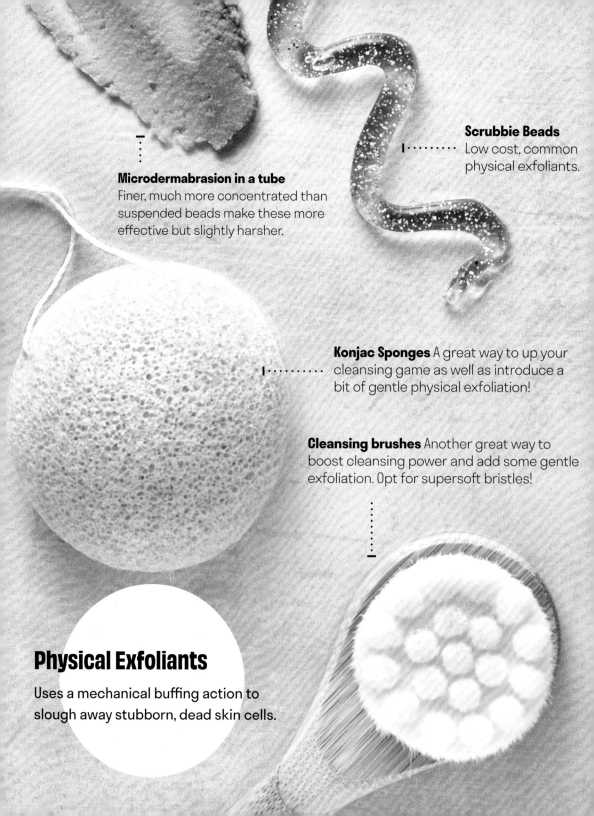

Microdermabrasion in a tube
Finer, much more concentrated than suspended beads make these more effective but slightly harsher.

Scrubbie Beads
Low cost, common physical exfoliants.

Konjac Sponges A great way to up your cleansing game as well as introduce a bit of gentle physical exfoliation!

Cleansing brushes Another great way to boost cleansing power and add some gentle exfoliation. Opt for supersoft bristles!

Physical Exfoliants

Uses a mechanical buffing action to slough away stubborn, dead skin cells.

Serums Usually in the 5% to 15% neighborhood. Great for daily upkeep..

Chemical Exfoliants

Chemically weakens bonds between overdue skin cells to help promote shedding.

Peel Pads Usually at a more intensive level. A convenient way to have a weekly peel!

Rinse-off Masks Can range from ultra gentle enzyme peels to intensive, high-dose glycolic peels.

Toners Usually very gentle. Suitable for those with sensitive skin.

A General Chemists' Guide to Choosing AHAs

How to level up your AHA game cautiously!

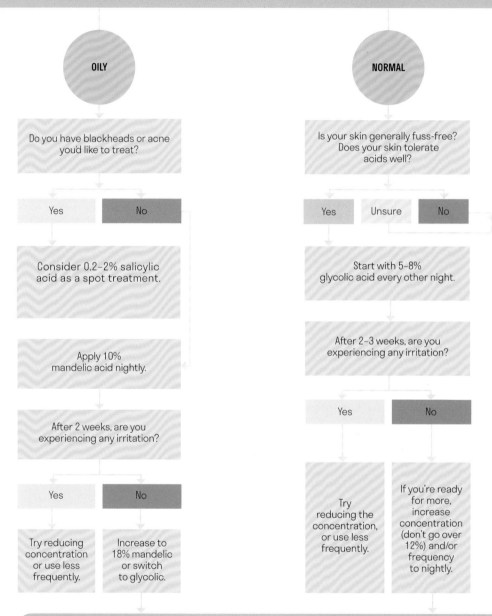

OILY

Do you have blackheads or acne you'd like to treat?

Yes	No

Consider 0.2–2% salicylic acid as a spot treatment.

Apply 10% mandelic acid nightly.

After 2 weeks, are you experiencing any irritation?

Yes	No

Try reducing concentration or use less frequently.

Increase to 18% mandelic or switch to glycolic.

NORMAL

Is your skin generally fuss-free? Does your skin tolerate acids well?

Yes	Unsure	No

Start with 5–8% glycolic acid every other night.

After 2–3 weeks, are you experiencing any irritation?

Yes	No

Try reducing the concentration, or use less frequently.

If you're ready for more, increase concentration (don't go over 12%) and/or frequency to nightly.

R E A D Y   T O   L E V E L   U P ?

DRY

Start with 5-10% lactic acid nightly.

After 2-3 weeks, are you experiencing any irritation?

Yes	No

Try reducing concentration or use less frequently.

If you're ready for more, switch to 5-10% glycolic acid nightly.

ADD A 20-30% GLYCOLIC OR LACTIC ACID MASK ONCE A WEEK.

Do you have sensitive skin?

While acids may sound intimidating, you too can reap the benefits. Let's just take it slow!

There are three scenarios you may find yourself in:

1. I'm experiencing tingling, stinging, dryness, and redness.

Okay! You've hit Irritation Town. Dial back all actives and focus on getting back healthy skin with the basic routine of cleanse, moisturize, and sunscreen.

2. My skin seems to be doing well!
Great—stay the course.

3. Nothing is happening.
Remember, it takes at least four weeks to see any noticeable difference. But it may be time to level up! Use the acid ladder below to help you choose your next AHA.

Start with 10-14% gluconolactone

10% mandelic acid

5-10% lactic acid

5-10% glycolic acid

* Remember, if at any point you find yourself encountering Scenario 1, stop all actives use and focus on getting skin back to its healthy self.

PERSONAL TAKE

By now, you can see that there's no one standard answer to skincare. It's all about your current skin situation and what works for you. Here are some of our personal notes on this category.

Gloria

AHAs are another one of my categories! My skin is highly acid-tolerant, so I have to say glycolic all the way. I do enjoy using mandelic acid as a gentler alternative if I'm using other aggressive active ingredients such as retinol. I personally feel very "meh" about lactic acid based on the stickiness and smell alone. I also enjoy using a very gentle physical exfoliant in addition to the chemical exfoliants to keep the glow going. BHAs, on the other hand ... those are tough for me. They're just too drying for my dry, dry, dry skin.

Chemical exfoliants have probably loitered around my skin routine all my life. Going through cystic acne as a student, I definitely went way overboard on the salicylic acid. What's worse is that glycolic acid also ended up in a few of my acne washes and toners. It's too easy to get carried away with this category when you want to get rid of that inconvenient pimple fast. The weird thing is, now my skin doesn't tolerate glycolic all that well. I now interchange as needed. During high-level retinol moments, I'll add in a little lactic and gluconolactone. For summer months, when my skin is more congested, I'll add in salicylic and mandelic.

Victoria

CHEMICAL EXFOLIANT FAQs

1 Q: This got really complicated! So, where should I start?

A: A good general starting point is 5% glycolic acid.

2 Q: What does a chemical burn do to your skin, anyway?

A: It'll feel a lot like a sunburn—red, stinging, itchy, just all-around irritated skin. In more serious cases, you end up with discolored, rough-textured skin.

3 Q: If you had to pick just one to always use, would it be physical exfoliation or chemical exfoliation?

A: Go chemical. These ingredients not only buff the stubborn cells on the surface but also come with extra benefits to brighten skin and reduce fine lines and wrinkles.

4 Q: So, if they're acids, how come they eat away at the bonds attached to cells that need to go, rather than cells that should still be there?

A: Great question! In a nutshell, they don't penetrate deep enough to eat away at the bonds that are supposed to be there. And this is how you overexfoliate, too. If you go too crazy on your chemical exfoliants, they will get to the cells that are supposed to be there. That's how you end up with irritated skin.

5 Q: Is there such thing as too much chemical exfoliation?

A: Of course! This still follows Chemists' Commandment #6. Some signs of going too far are flaking, prolonged stinging, and redness.

RETINOIDS

Retinoids, or vitamin A derivatives, really function as the *A* in the ABCs of active skincare. This whole class of ingredients has been considered skincare royalty since its discovery, treating everything from acne and hyperpigmentation to lines and wrinkles. However, retinoids do come with a few trade-offs. They're also often known for their side effects of redness, flaking, and stinging. But with proper use, most of these retinol side effects can be toned down to a minimum so you can reap all of those anti-aging benefits. In this section, we'll break down the category, going through the differences among the types of retinoids found in stores. Then we'll help you determine the best retinoid for your skin needs and how much you should use to maximize your retinoid experience with minimal irritation.

WHY RETINOIDS?

Retinoids have one of the longest histories of testing and have been found to treat acne, wrinkles, and pigmentation. Consider this category the gold standard when it comes to wrinkle fighting because of its ability to promote collagen production and prevent collagen degradation. How? So glad you asked!

Once Upon a Collagen

Before we get into all the good things retinoids do for your skin, we have to talk about the story of collagen. When you think of anti-aging, collagen is probably a term you've seen floating around on the labels of your wrinkle-fighting creams. That's because collagen—or, more precisely, the loss of collagen—is a key culprit in chronoaging.

Collagen actually makes up roughly 75% of the dry weight of your skin. With its tight-knit, triple-helix structure, it is responsible for the skin's overall structural integrity. As such, the loss of collagen over time is a root cause of wrinkles and sagging.

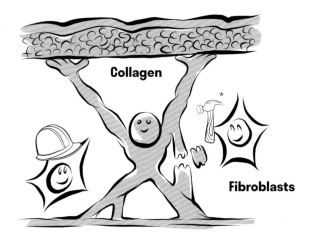

FIBROBLASTS ARE THE CELLS RESPONSIBLE FOR MAKING COLLAGEN

* Fibroblasts do not actually work with hammers.

Enter fibroblasts: These cells not only play a critical role in wound healing, they also secrete procollagenase, a precursor that acts as the building blocks that are cross-linked to make the final product, collagen.

A Very Brief History of Retinoids

You can actually trace retinoids all the way back to ancient Egypt, where vitamin A–rich liver was eaten to treat night blindness. Retinol itself dates back to World War I, when research showed that vitamin A deficiency caused dry skin and keratosis pilaris. However, it wasn't until the late 1960s when Dr. Albert Kligman, Dr. James Fulton, and Dr. Gerd Plewig invented tretinoin. Dr. Kligman was a controversial but prolific pioneer in the field of dermatology and the first to debunk the myth of chocolate causing acne, studying the stages of acne, and coining the term "photoaging."

Have a Talk with Your Doc

The "clean beauty" movement has led to some confusion on what ingredients are actually safe. If you dig around on the interwebs, you'll find a few brands swearing off retinol because it's deemed "dangerous." This stems from some studies saying that excess vitamin A during pregnancy can be teratogenic (cause birth defects.) Thus, doctors won't prescribe oral Accutane when you're trying to get pregnant. Topical retinoids pose much less of a risk, but doctors will err on the side of caution and typically suggest avoiding retinoids in these cases. So, if you're pregnant or thinking of trying, first consult your doctor on what skincare ingredients to avoid. Otherwise, retinoids are perfectly safe!

BUT SERIOUSLY, WHY RETINOIDS

Collagen doesn't live forever. The most abundant collagen in your skin, collagen 1, has a lifespan of about 30 years. This presents a unique challenge, because as you age, a buildup of modifications and cross-links occurs with the old collagen, leading to collagen fragmentation (aka wrinkles, sagging, and everything that will just bum you out.) This can be aggravated by the accumulation of sun damage, but also just by time itself. Even worse, it's difficult for fibroblasts to completely remove these modifications, and they can't be incorporated into new collagen, disrupting the structural integrity of the dermis. Additionally, with collagen fragmentation, there's a loss of sites where the fibroblast can attach, stretch out, and happily signal for procollagen production. Thus, collagen fragmentation is another root cause of our chronoaging (natural-aging) wrinkle worries.

Retinoids and Your Collagen (Plus Other Skin Benefits!)

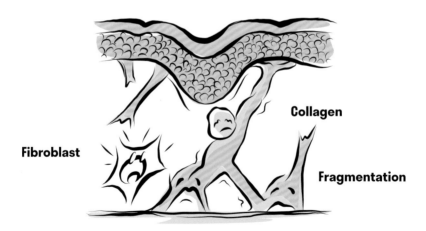

Fibroblast

Collagen

Fragmentation

COLLAGEN FRAGMENTATION MAKES IT HARDER FOR FIBROBLASTS TO REPLACE OLD COLLAGEN, LEADING TO STRUCTURAL PROBLEMS (AKA WRINKLES)

Do not lose hope yet! The good news is that the retinoids are here to save the day. Retinoids interact with your skin's retinoic acid receptors (RARs) to stimulate collagen production, improve fibroblast proliferation, *and* prevent collagen degradation. Of the many anti-aging ingredients out there, retinoids are some of the only ingredients proven to cover the full spectrum of collagen protection and promotion.

Speaking of full spectrum, we also cannot forget that this category is also backed by a plethora of studies in treating severe acne and post-acne pigmentation, too. In fact, many people's first experience with retinoids may come from acne treatments.

"Wait a minute," you might say. "It sounds like retinoids literally do everything. Sign me up!" Well, nothing is perfect. Using retinoids does come with some challenges and annoyances. For the most part, this category of ingredients is not the most stable bunch and easily degrades (read: more headaches for chemists!). There are also the notorious side effects of peeling, sensitivity, and redness, which have deterred many. Finally, not all retinoids are alike, but with a little bit of guidance, you'll find this to be an active ingredient you can use throughout your lifetime.

DEBUNKED:
Don't Be So Thin-Skinned

Shedding is a hallmark side effect of retinol, and may be one of the reasons why some people are worried about trying it out. There are even fear-mongering articles out there that suggest retinol thins out your skin permanently. Some might even say retinol damages your skin barrier function. Although it's a bit counterintuitive, consistent retinol use actually thickens both your dermis and epidermis—leading to better hydration and skin barrier function. So, hang in there! After the initial "retinization hump," it's all good things from there.

MEET THE RETINOIDS

N ow that we're four pages into this chapter, we probably should define what *retinoids* mean. The term *retinoids* is actually an umbrella term for all the vitamin A ingredients you can score on the market. For example, retinoic acid, retinol, and retinyl palmitate are all part of the bigger retinoids family.

But with all of the retinoids on the market, how can we keep track of them all? You can generally break them into three classes: prescription, over the counter (OTC), and cosmetic.

Prescription: Let's visit the dermatologist!

	Trade Name	%	Recommended for	Benefit
Isotretinoin	Accutane	N/A	Severe cystic acne	Oral medication. Not typically the first mode of treatment prescribed.
Tazarotene	Tazorac 0.1%		Acne, post-acne hyperpigmentation	Shows solid benefits treating postinflammatory hyperpigmentation and acne in darker skin tones.
Tretinoin	Retin-A	0.01–0.1%	Acne, wrinkles	The gold-standard retinoid; the longest history of skin benefits.
Adapalene	Adapalene	0.3%	Acne	A newer generation retinoid that shows great data treating acne with less skin irritation.

Listed above are a few commonly prescribed retinoids. To obtain any of these, you'll need to take a trip to the dermatologist's office. These retinoids are typically used to treat moderate to severe acne. We highly recommend that if you are struggling to manage your acne breakouts, it's time to seek out a good dermatologist. Derms will be able to diagnose your acne and come up with a tailored course of treatment. We truly believe a derm partnership will be essential if you want to get serious about tackling your acne.

Over the counter (OTC): Let's hit the pharmacy!

	Trade Name	%	Recommended for	Benefit
Adapalene	Adapalene	0.1%	Acne, hyperpigmentation	Gentler than tretinoin or retinol, with solid data about treating acne.

Adapalene recently became available as an OTC ingredient, and we couldn't be more excited. A newer member of the retinoid family, this synthetic retinoid has been shown to be more gentle than tretinoin, but still effective in treating mild to moderate acne. There's even data that suggests it can help with post-acne marks. Use an adapalene gel before your moisturizer step. Even though it is more gentle than tretinoin, we still recommend pairing this with a good, soothing moisturizer to keep any added dryness or potential irritation to a minimum.

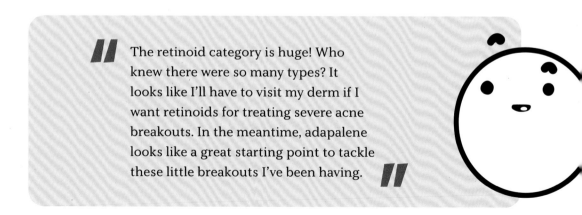

The retinoid category is huge! Who knew there were so many types? It looks like I'll have to visit my derm if I want retinoids for treating severe acne breakouts. In the meantime, adapalene looks like a great starting point to tackle these little breakouts I've been having.

In stores: Let's go shopping!

	Also known as	%	Recommended for	Benefit
Retinol	Retinol	0.1–1.0%	Wrinkles, hyperpigmentation	The gold-standard retinoid that you can buy in stores. Very effective and commonly found.
Retinaldehyde	Retinal	0.05–0.15%	Wrinkles	Not as easy to find in products because of serious stability issue. But an alternative to retinol if you're looking for something with slightly more efficacy.
Retinyl Palmitate	Retinyl palmitate	~1%	The trash can. Really, this is pretty pointless.	We find nothing beneficial about this ingredient. (See the next section for more info on why!)
HPR	Hydroxypinacolone retinoate, granactive retinoid	~1%	Sensitive skin types looking for something more gentle	New kid on the block with not a lot of data available; recent findings show that this gentle alternative also provides anti-wrinkle benefits.
Bakuchiol	Marketing sometimes refers to this as "plant retinol"	0.5–1%	Sensitive skin types, acne, wrinkles, hyperpigmentation	New kid on the block, also known as a "plant alternative" to retinol. Not a lot of data available.

Since neither retinol nor retinaldehyde is the most stable of actives to work with, many companies have developed different versions to achieve three main goals: more stability, less irritation, and prolonged efficacy. Of these new-gen retinoids, the ones that have caught the most traction are hydroxypinacolone retinoate (HPR) and bakuchiol (a plant-based ingredient). Our chart shows preliminary concentrations for tested effective levels, but we should mention that for any new ingredient like these two, there's still a lot of data required to really understand how these actives work and the mechanism that's giving those skin benefits. While the jury is still out, we do recommend these alternatives for those who have really struggled with retinol and retinal.

Between retinol and retinal, retinal is actually the more potent of the two. Sadly, it's very difficult to find because retinal is annoyingly even more unstable than retinol. Thus, retinol is truly the gold-standard retinoid that you can readily buy at the cosmetic level, and the data does support it. It has a long history of data showing its ability to tackle wrinkles and even hyperpigmentation, but it's one of the fussier actives to incorporate. It also comes with those side effects of redness, flaking, dryness, and stinging, so it's important to add this into your routine slowly and start at a low concentration. Once you have a good idea of how your skin behaves on retinol, you'll be able to reap all the benefits. How do you do this? So glad you asked! Let's prepare you for your retinol journey!

INGREDIENT HIGHLIGHT #2:
Retinyl Palmitate

This stuff is basically retinoic and palmitic acid bound together. This is actually the dormant form that your skin naturally stores when there's excess vitamin A. Retinyl palmitate is appealing because it's slightly more stable than retinol, and is pretty darn easy to incorporate into formulas. But that's the end of the good news for this molecule. It's hard for retinyl palmitate to bring that retinoid fighting power, since its skin-penetration ability is questionable. To make matters worse, its journey to the retinoid receptor is much longer: It must transform from retinyl palmitate to retinol to retinaldehyde to retinoic acid. This is why retinyl palmitate is considered the weakest retinoid on the market.

Oh! I'm awake.
Time to become retinoic acid.

Retinyl palmitate is the precursor to retinol. It has no real biological activity, aside from being the storage form for excess retinol. Consider this guy the least effective of the bunch.

SHOPPING FOR RETINOL

Don't let the phrase "cosmetic grade" fool you—retinol is a potent molecule. But its place in the cosmetic section means that it takes some detective work to discern winning products from questionable ones. One of the most important considerations is that retinol is not the most stable molecule. (It's really a big problem for the whole retinoid family.) Retinol is sensitive to light, air, and heat. Exposing retinol formulas to the elements can lead to an untimely retinol death before it can even deliver its benefits to your skin.

Packaging

The package should be a big deciding factor in your search for retinol products, as it can help sustain the product's shelf life by limiting the amount of air it's being exposed to and shielding the formula from light. Aluminum tubes with a tiny nozzle are the ideal packaging, with the least amount of oxygen exposure. Airless pumps are the next best in terms of protecting the formula. Dropper bottles are pretty "meh." They can help minimize light, but you still expose the formula to air every time you open it. And just say no to jars.

RETINOL PACKAGING MATTERS (AIR EXPOSURE MEANS RETINOL DEATH)

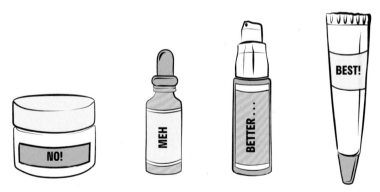

Formula Consideration

You can probably tell by this point in the book that "Concentration is key!" is one of our favorite Chemists' Commandments. With a potent active like retinol, it's even more true. Opt for a product that tells you how much retinol it uses. (We recommended between 0.1% and 1%.) Also, be on the lookout for good supporting ingredients. Another chemist-favorite active, niacinamide, has been shown to reduce the irritation of retinol. We love products that combine the two!

Storage

For any active ingredient that isn't stable, think "vampire" settings: Store your retinoids in cool, dark places out of the sunlight to extend shelf life.

Use

We recommend between 0.1% and 1%. The most important thing here is how to use it while minimizing the pain of shedding, sensitivity, and redness. The general approach is to go slow: If you're a beginner, start with a 0.1–0.3% dose of retinol. Try to use it two or three times a week, and monitor how skin reacts. If there's minor flaking or redness, but no lingering stinging, stay the course. Remember that everyone's skin is different and thus tolerates ingredients differently. Some people will stay on a 0.1% dose for years before starting to work their way up.

CHEMIST CONFESSIONS: SHOPPING RED FLAG

In the Wild West that is Amazon.com, you can find a great deal of ... well, questionable beauty products. We have found some that have claimed a whopping concentration of 2% retinol. So, what's the deal? Well, either they're fibbing and the product is actually 2% of a retinol-containing raw material (and the actual percentage is much lower), or they're telling the truth and there really is that much retinol in the product. In the case of the former, they deserve an eye roll; if it's the latter, abort! That's way too much retinol! Let's not bring that kind of irritation to the face. Honestly, neither of these scenarios deserves your business!

TROUBLESHOOTING YOUR ROUTINE

Just like all the other actives, it's possible retinol might not be a good fit. There will be a select few of you who, even at a 0.1% concentration, will continuously struggle with the irritation. That's just the way it goes; skin has a sense of humor like that. To help you troubleshoot your retinoid situation, here are a few checkpoints you can consider:

 The irritation *is* irritating! You've been struggling for a long, *long* time with the shedding, sensitivity, and redness, and you're only at 0.1%. It might be time for a switch. The good news is that there are now more gentle alternatives, like bakuchiol and hydroxypinacolone retinoate, that you can also consider.

 My acne is angry! For all acne struggles, we prefer adapalene. This synthetic retinoid is now available as an affordable, OTC topical. There are several strong studies that have shown 0.1% adapalene performing on par with low amounts of tretinoin.

 I can't be flaking right now! For those big moments, it's okay to give retinoids a rest. Plan one to two weeks out, before that big event, to pause the retinol so your skin will be flake-free and makeup ready.

> **"** Minimize other potential irritants like acids when starting your retinoid journey. Once your skin has a handle on things, taking a rest day and inserting a gentle chemical exfoliant into your routine can help minimize the excess flaking. **"**

CHEAT SHEET
Retinoids Summarized

Chemists' Guidelines:

- Retinol is one of the main retinoids you can purchase in stores that can provide hyperpigmentation help.

- Tretinoin is a prescribed topical that's commonly used for acne treatment and is the gold-standard retinoid. Seek out a derm.

Recommended Retinol Starting Point:

Beginner Start in the 0.1–0.3% concentration range. Use 2–3 times a week until your skin has acclimated.

Expert You're working your way up to 0.5–1.0%. Still consider starting out at 2–3 times a week until skin has acclimated.

Sensitive Sometimes it's just not meant to be; look into alternatives such as hydroxypinacolone retinoate or bakuchiol.

Acne Seek out OTC synthetic retinoid adapalene in drugstores.

- Start slowly, with just 2–3 applications a week. Typical acclimation takes anywhere from one to six months. Patience is key!

- If you have a good handle on your skin's irritation moments with retinoids, you've mastered your retinoid routine.

PERSONAL TAKE

Gloria

Oof—this is a tough one for me. As tolerant as my skin is with acids, I turn into a sensitive, shedding snowflake as soon as retinol touches my skin. I am talking about full-on snake molting with just 0.3% retinol. It has been my 2024 skin resolution to figure out my magic routine, one that that incorporates retinol to level up my anti-aging game. By the time this book is out, I hope to have reined in this bad boy.

I love retinoids! Bring on the shedding, because this category has been a staple since college, when I had major cystic acne. As I head into my thirties, my skin's tolerance for retinoids has changed, so I've had to tweak how many times I use them, but bring on the 1%! Overall, I've been happy with the results. Retinols not only help manage my breakouts, but there are also definite visible improvements in fading those post-acne spots. My best advice is to just be patient. If you can get through the initial hump, it can be a really gratifying active to have in your arsenal.

Victoria

RETINOID FAQS

1

Q: How should I layer retinoids?

A: Retinoids are usually in cream or oil form. So, they come after your serum and before your moisturizer. Some retinol formulas are even moisturizing enough to replace your night moisturizer.

2

Q: I've been using retinoids for 10-plus years. Any downside?

A: Nope! You might notice your skin-sensitivity levels changing and may need to adjust your use frequency or take breaks once in awhile. But there isn't a long-term downside to using them. The one time you should take a break from retinoids is when you're pregnant.

3

Q: I switched brands but used the same 0.5% retinol concentration, and my skin reacted differently. Why is that?

A: Because retinol has become so popular, chemists are hard at work formulating new and improved formulas that can include things like soothing ingredients or the latest encapsulation technology. This can explain why switching to a different product with the same concentration can result in a different reaction. Just remember to give the new product at least 8 weeks to see how skin responds.

4

Q: I want to use retinoids but don't have acne; is there anything I should keep in mind when I choose a product?

A: If your skin's on the dryer side and you want to start retinoids, pair with a moisturizer with higher occlusives, since a common initial side effect is flaky dryness.

5

Q: I heard retinol is unstable. When should I throw out my retinol?

A: Responsibly formulated and packaged retinol will still last up to one year after opening. (Check the package label.) Make sure that you store it in proper vampire conditions! Two telltale signs of degraded retinol are a change in color and oil seeping out of the packaging.

VITAMIN C

Vitamin C products are so prolific in the skincare industry, it may feel like brands are beating a dead horse here. But the dead horse works! In fact, it's quite the skincare multitasker. Vitamin C has decades of clinical testing to prove that it is an effective free radical quencher, collagen booster, and skin brightener. Of course, things aren't as simple as they appear. Did you know that there are many types of vitamin C ingredients in skincare? And not all of these are created equal? Did you know that the good vitamin C serums can actually smell like hot dogs? Well, let's dive in to take a closer look at these chemists' favorite molecule!

WHY VITAMIN C

Vitamin C works on skin in three major ways: age prevention, tackling uneven skin tone, and lessening fine lines and wrinkles. It all boils down to the fact that vitamin C is an incredibly effective topical antioxidant.

Antioxidants vs. Free Radicals

The term *antioxidant* (AOX) has been almost abused in the beauty and food industries to the point that most people don't understand what it actually means. Antioxidants are ingredients that prevent oxidative damage from free radicals. Sound familiar? If you refer to the sunscreens chapter (starting on page 70) you may remember the free radical. The thing is, free radicals can come from anywhere. They even play an important part in our own natural cell functions. But in the context of skin, free radicals are highly reactive molecules generated by sun exposure, infrared A, cigarette smoke, stress, and so on. To explain free radical damage, how about a story?

STORY OF A FREE RADICAL

UV rays generate free radicals	Free radicals wreak havoc on cells	Free radical finds love with the AOX and leaves cells in peace

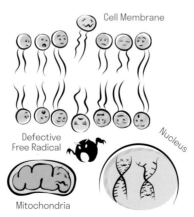

Cell Membrane

Defective Free Radical

Nucleus

Mitochondria

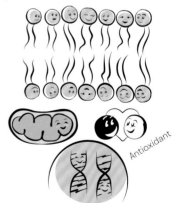

Antioxidant

Once upon a time, too much sun caused free radicals to generate in the skin. Each one of these very unstable free radicals was trying to find a way to reunite with its missing electron. They wreaked havoc upon the skin cells and their major components, the nuclei, mitochondria, and cell membrane. It was only when the free radical found the antioxidant that the free radical was quenched and there was peace—among the skin cells and across the skin universe.

An Orange a Day Keeps Scurvy Away

Fun fact! Long before vitamin C was used for your youthful glow, voyagers in the 1700s realized that a diet with sufficient fruit prevented scurvy. In fact, the Latin name of scurvy, scorbutus, is how ascorbic acid got its name. Finally, in the early 1900s, vitamin C was identified, isolated, and synthesized. It wasn't until the early 1990s, however, that vitamin C started entering the skincare scene. Then, in 1992, Dr. Sheldon Pinnell, founder of Skinceuticals, pioneered and concocted the famed CE ferulic combination, earning an iconic (and now-expired) patent. Since then, the vitamin C serum has become a staple in many people's anti-aging skincare routines.

Vitamin C Around the World

Japan has a class of products called quasi-drugs. In our minds, this category came about because someone said: "It's *sorta* a drug in terms of functionality, but we really don't feel like dealing with the amount of paperwork that goes into approving something like Viagra." What that means for us as consumers is that ingredients that fall under this category have more testing and verified efficacy than the average cosmetic product. In Japan, whitening products are very popular and fall under this "quasi-drug" category. Vitamin C and some derivatives are also notable ingredients under this category because of how effective they are. So, for all you skeptics, this ingredient category is Japan-approved!

VITAMIN C: SKINCARE'S GOLD-STANDARD ANTIOXIDANT

Our skin has its own natural antioxidants: CoQ10, glutathione peroxidase, superoxide dismutase, vitamin E, uric acid, carotenoids, melanin, and vitamin C. Sound familiar? That's right—you can find all of these in skincare too. Naturally, you would think that anything found in skin must be good for your skincare regimen, but that's not actually how it works. All of these actives need to be tested to show that they are stable in a topical formula and can deliver the intended skin benefits. This is why L-ascorbic acid, or L-AA (pure vitamin C) is the reigning, gold-standard skincare antioxidant. It's also one of the rare few that has been proven to work topically as an antioxidant.

Besides being an antioxidant that helps with age prevention, vitamin C is a fantastic skin-brightening ingredient. The almighty C has also been demonstrated to treat melasma and hyperpigmentation effectively, either alone or in conjunction with other brightening actives. If you're curious to learn more about how to put together a hyperpigmentation-fighting routine, you can find detailed advice in our Routines chapter, starting on page 188.

// Fun fact! What do humans have in common with guinea pigs? We're two of the few mammals that can't manufacture our own vitamin C. This is why a diet rich in vitamin C is so important. You're welcome, trivia nerds!

Last, this multitalented molecule has even been shown to boost collagen synthesis. (Refer to pages 132–134 for a refresher on why you should care about collagen.) Very few ingredients actually have been proven to boost collagen, and since collagens are the pillars of your skin's structural integrity, this explains why ascorbic acid also helps with minimizing fine lines and wrinkles.

In a nutshell: The three main benefits L-AA offer are photoprotection, brightening skin tone, and boosting collagen. With this trio of benefits, L-ascorbic acid can be considered both a preventative and a direct anti-aging fighter. It can deliver both short-term, visible improvements, like improved skin tone and clarity, and long-term anti-aging benefits.

Rubbish Alert!

Can a DIY lemon-juice concoction give me those sweet, sweet vitamin C benefits?

So, to answer the question "Can lemon juice be used as a vitamin C serum?" let's first look at the composition of lemon juice:

> pH: about 2.2
> Citric Acid: 5%–8%
> Ascorbic Acid: about 0.7g/mL of juice, or 0.7% by weight

Most good, reliable vitamin C serums you can buy will have at least 5%–15% vitamin C. So ... from the composition, we can conclude that using neat lemon juice, with a paltry 0.7% ascorbic acid, doesn't provide nearly the same level of vitamin C as a well-formulated serum found in stores. So, let's just leave the lemon juice for tea and fish. Lemon juice can effectively "acidify" your acid mantle, but that's about it. The reality is, the pathway that something uses to get into your body matters. So, while lemon juice may be a good source of vitamin C from a nutritional standpoint, it doesn't really translate to good topical use.

MEET THE VITAMIN C FAMILY

 Pro Tip: Before we get started, here's a quick "decode that ingredient label" tip for everyone: Not sure what ingredient is the vitamin C? Look for anything with *ascorb* in the name.

Vitamin C is actually the broad umbrella term for a whole family of molecules. All these awesome benefits we just described? We're actually just talking about ascorbic acid (L-AA). There are actually many more vitamin C derivatives you can get in skincare. So far, these derivatives fall short of ascorbic acid in one way or another. However! A few of them are promising and can be great choices for those looking for a more stable, gentle option. If you're curious about the world of vitamin C derivatives, we gotchu.

We'll differentiate these derivatives according to what we'll call the Vitamin C Trio (of benefits): brightening, photoprotection, and boosting collagen. Now, there are a lot more benefits that aren't listed here, but these are some of the more common ones you'll encounter as you're shopping for your next vitamin C treatment. Because L-AA sits at a lower pH, it isn't for everyone, even at a low dose of 5%. Use this chart to help find a derivative replacement to sub in and fill the role. Happy shopping!

Meet vitamin C's best pals: vitamin E and ferulic acid

Have you ever wondered why vitamin C is paired with vitamin E and ferulic acid? Vitamin E (ingredient name tocopherol) and ferulic acid are both antioxidants as well. The trio works together synergistically. Including vitamin E and ferulic acid in your serum along with vitamin C not only slows down vitamin C degradation, but it also enhances the free radical–fighting power of the entire formula. In addition to these positive side effects, there's the slightly weird side effect of smelling … meaty. No, it's not your imagination. Who knew? The Fountain of Youth smells like hot dogs.

	L-Ascorbic Acid	Magnesium Ascorbyl Phosphate	Sodium Ascorbyl Phosphate	Ascorbyl Glucoside	3-o Ethyl Ascorbic Acid	Ascorbyl Palmitate
Description	Chemist Nickname: L-AA The gold-standard vitamin C in terms of performance. If you find your skin to be easily irritated by L-AA, consider other vitamin C alternatives.	Chemist Nickname: MAP Data shows potential tackling melasma.	Chemist Nickname: SAP Data shows potential tackling acne.	Chemist Nickname: AA2G Popular in Japanese and Korean products.	Chemist Nickname: Et VitC Data is pretty spotty for this one.	Chemist Nickname: Garbage! (Just kidding.)
Target pH	Under 3.5	7	7	Doesn't matter! It's an easygoing guy.	5.5	Under 6
Effective %	Between 5–20%	At least 5%	At least 5%	At least 2%	At least 2%	Between 5–20%
What It's Good For	Does pretty much everything! Truly the gold standard in the vitamin C Trio of functions (brighten, photoprotect, boost collagen).	A support ingredient in your pigmenta-tion-fighting routine.	It's a bit sub-par in all three core vitamin C functions, but does come with bonus acne-fighting data.	Fighting hyper-pigmetation with potential collagen-boosting benefits.	Generally positioned as an ingredient to combat hyperpigmen-tation.	Good question! Kinda useless. It's pretty much good for … nothing.
What It's Not Good For	We meant it— gold standard! Like, what *isn't* it good for?	Anything on its own. There really isn't any data that shows that this packs enough oomph to be a stand-alone active.	It's weakest in the colla-gen-boosting department. We wouldn't recommend this molecule for mature skin.	Not so good as an antioxidant.	We would only recommend this if you can't get your hands on any other forms of vitamin C.	You should walk away from any vitamin C product if it only contains ascorbyl palmitate.

IN PRACTICE: L-ASCORBIC ACID

This formulation is notoriously unstable. It's very susceptible to the degradation trio: sunlight, water, and oxygen. Not to mention that temperature speeds up the whole process of ascorbic acid death. So, what that means for us chemists is that we have to get slightly creative with vitamin C formulas. This is why you can find a variety of ascorbic acid product types on the market. Here's a quick breakdown!

Classic Serum

How to identify in the wild: Take a quick look at the ingredient list. Is the first ingredient water? Then yes! It's a classic water-based serum.

Pro: Easy onboarding. It's a no-brainer! Use as you would any water-based serum. This is the format we would recommend for most people.

Con: Shelf life is meh. Even the classic CE ferulic formula can't keep vitamin C stable if you don't protect it with proper storage conditions. Store in dark, cool areas and make sure to properly close the container every time. Oh, and did we mention it smells like hot-dog water . . .

Pro tip: Given that this is a water-based serum, you should incorporate it into your first step post-cleansing.

Silicone-/Oil-Based Serum

How to identify in the wild: Is the first ingredient not water? Do you spy a bunch of oils like squalane or silicone oils? Then it's most likely an oil-based serum.

Pro: It's more stable than your classic water-based serums. Instead of dissolved vitamin C, it's actually crushed, *suuuuper* fine particles of L-AA suspended in the oil of choice.

Con: This texture isn't the best. It's an acquired taste. It can often be gritty and feel a little too . . . oily.

Pro tip: Okay. So, we know we just said that vitamin C serums should be the first step after cleansing. This is the exception. If you use another toner or hydrating serum, you should use this after the toners or serum step. You don't want the heavy silicones in these products getting in the way of the other good ingredients!

Powder

How to identify in the wild: Well . . . it's a powder.

Pro: This is by far the most stable product category. It's also relatively cheap based on the sheer amount of active ingredient you get.

Con: It's really messy! We've definitely had moments when we leave what looks like an illicit drug problem on our counters thanks to vitamin C powder.

Pro tip: L-AA is on Team Water, so, it's best to dissolve these powders into your toner or serum to maximize their effectiveness. Make sure you give the ingredient list a quick scan! We have seen "vitamin C powders" that actually use non-L-AA forms of vitamin C. (Seriously, what is the point of doing this?)

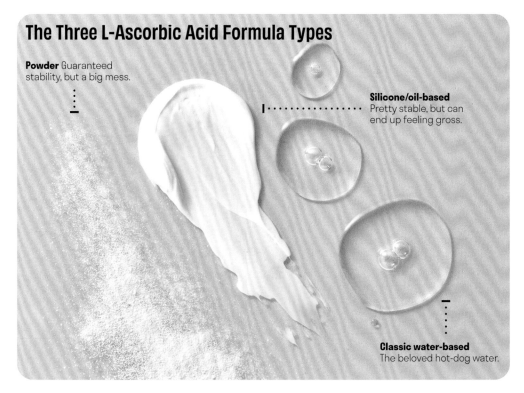

The Three L-Ascorbic Acid Formula Types

Powder Guaranteed stability, but a big mess.

Silicone/oil-based Pretty stable, but can end up feeling gross.

Classic water-based The beloved hot-dog water.

VITAMIN C PRODUCT STABILITY

Regardless of the product type you choose, stability is the most important thing to keep an eye out for. We're all guilty of leaving products in a forgotten corner until we rediscover them during spring cleaning. Vitamin C is a product you want to use through as soon as you open it. Clear, faint yellow should be your starting color. Even when it gets slightly orange, it's still functional—but it's a cue for you to hurry up and finish. If it matches the color of fine whiskey? Time to move on.

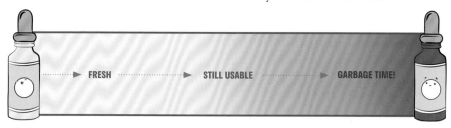

FRESH — STILL USABLE — GARBAGE TIME!

Vitamin C Troubleshooting

Not sure how to start, or feeling a bit lost on your vitamin C journey? Consider these guidelines:

Where does it go in the routine? Vitamin C should be your first step post-cleansing in the mornings—before your moisturizer and sunscreen. You can also use it at night if you'd like!

Just starting out? Start with a simple 5% ascorbic acid serum

Ready to level up? Try the classic CE ferulic combination with 15% L-AA.

Is 15% too irritating? No worries! A 5% L-AA serum is still an effective concentration. If that's still irritating, it's time to scope out an alternative vitamin C molecule.

> **Is your top skin concern anti-aging?** We recommend going to sodium ascorbyl phosphate next.

> **Is your top skin concern pigmentation?** We recommend trying ascorbyl glucoside.

CHEAT SHEET
Vitamin C Summarized

Chemists' Guidelines:

- L-ascorbic acid, or L-AA, is the gold standard, with proven data as an antioxidant, wrinkle fighter, and pigmentation fighter.

- Despite 20% ascorbic acid serums being all the rage, you only need 5% to start taking advantage of the benefits.

- Ascorbic acid serums are the perfect pairing with your sunscreen, but they can also serve as a night serum as well.

Recommended Starting Point:

Beginner   Start with a 5% ascorbic acid serum.

Expert   15–20% ascorbic acid serums are the way to go.

Sensitive   Sometimes it's just not meant to be. Look into alternatives such as sodium ascorbyl phosphate and ascorbyl glucoside.

Chemists' Hacks and Pro Tips:

- Ascorbic acid serums have a pretty crummy shelf life. A faint yellow is okay, but once a serum starts browning like a banana worthy of banana bread, it's time to let it go.

- Store in vampire settings! Cool, dark places, people.

- Doing a treatment? Be aware that your skin might be a little sensitive from a heavy-duty treatment session and you may need to take a break from vitamin C.

PERSONAL TAKE

Gloria

Vitamin C is one of my favorites in the Big Four list! My skin is incredibly ascorbic-acid tolerant. I have probably tried every other ascorbic acid product under the sun at various concentrations and pH levels without irritation. While ascorbic acid alone isn't super effective against my hyperpigmentation spots, there's a very distinct difference in that glow factor when I'm on an ascorbic acid serum. To me, my favorite form is actually powder vitamin C, just because I'm not a huge fan of layering 845,783 products. But my bathroom counter definitely looks a bit dodgy from the powder mess . . .

So, L-AA is one thing that feels like it might not be doing anything for me at the moment. I also even have moments when I'm on retinoids and my skin can be a little sensitive to L-AA. But I also want to tout that I will still use this religiously as an anti-aging product. I'll be in it for the long term, in the hope that my collagen will hold out just a little longer for me.

Victoria

VITAMIN C FAQs

1

Q: So, will eating citrus fruits also help my skin?

A: While eating citrus is a great way for your body to get some vitamin C, it won't be the most effective way to get those skin benefits we've been discussing.

2

Q: Will putting citrus on my face help my skin?

A: Ahem—see the Rubbish Alert section on page 151.

3

Q: I've heard so many good things about vitamin C, but my skin seems irritated and breaks out when I use it. What am I doing wrong?

A: It could be a case of "too much of a good thing." Nowadays, you can find serums with 20%, 25%, even 30% L-AA, which is really high! Try a product with less vitamin C, or a derivative, or even diluting it into your moisturizer. If you're still experiencing irritation, it could be that vitamin C is not right for your skin quirk. Check out page 179 for some other antioxidant favorites instead!

4

Q: Are there antioxidant-rich home ingredients that can help your face? Like face scrubs with chocolate?

A: Not ... really. The reality is, the concentration you'd need (about 10%) is super high for food. So, nothing in the pantry will come close to that level.

5

Q: So, if it's called ascorbic acid, is it going to feel like or work like other acid-derived products?

A: Nope! L-AA doesn't work the same way as other acids, such as glycolic acid.

NIACINAMIDE

Quick! Take a look at the ingredient lists of all the skincare products you own. It will probably surprise you how many times niacinamide might make an appearance in your routine. Niacinamide, or vitamin B3, is really the jack-of-all-trades when it comes to skin benefits. It's been quietly playing the supporting role for a few decades, but in recent years it's finally been getting the limelight it deserves. Although all the mechanisms of niacinamide are not completely ironed out, it has been found to help with oil control, strengthen the skin barrier, and even combat hyperpigmentation. So, let's learn all about the ultimate skin-active chameleon—niacinamide.

WHY NIACINAMIDE

Niacinamide is a form of vitamin B3 that also happens to be pretty versatile. Quick chemist confession! Because it can be easily incorporated into formulas, this active ingredient often gets thrown into all types of products as an "efficacy insurance policy." With its recent rise in popularity, you can find it in all sorts of products: serums, treatments, moisturizers, and even body lotions—which is great, since niacinamide has quite the résumé of benefits. But because it's in so many products, people are quite confused about what niacinamide is actually for. So, let's give it the proper love and attention it deserves by first figuring out what the hell it actually does.

How Does It Work?

The mechanism is not well understood, but there are some sweet findings: You'll notice that the unifying theme across these chapters is that the chemists' favorite actives are all multifunctional. Niacinamide, of course, is no exception.

Unfortunately, niacinamide's "biology" section is slightly lacking compared to other categories of actives. The reality is, the science gets pretty complicated. Niacinamide is directly tied to the metabolism. (Anyone remember the Calvin cycle from high-school chemistry?) Without getting into the nitty-gritty, this means that it happens to be linked to many biological pathways. Other than what's mentioned here, it's also been studied as a glycation fighter.

What's glycation, you ask? It's a reaction that involves free sugars leading to resulting advanced glycation ends (AGEs). And, yes, that acronym is just as foreboding as it sounds, since these AGEs can impact and impair skin-tissue elasticity, along with blood vessels, tendons, and the progression of certain diseases.

It promotes a healthy skin barrier and improves skin texture: Studies have shown that creams with as little as 2% niacinamide can improve skin moisture and overall skin barrier function in the long run. This is great news for those of you with dry skin! Better skin barrier function means your skin is better at retaining moisture, even in those harsh winter months. The general theory is that niacinamide promotes keratin synthesis and the stimulation of ceramides, both significant contributors to a healthy skin barrier.

It's an oily-skin regulator: Oil control can be a very frustrating issue for people, since almost nothing works long term. In fact, most oil-control products only mask this problem short term with oil-absorbing powders. Yet, good news! Niacinamide is actually one of the few ingredients out there that has demonstrated long-term sebum-regulation effects when used consistently

AKA Vitamin PP

Niacinamide has also been known as vitamin PP, for "pellagra preventative." It was identified in the 1920s as the crucial nutrient that prevents, well, pellagra. You probably don't even know what pellagra is thanks to the fact that you now get B3 in your diet. It wasn't until the early 2000s that researchers started paying attention to niacinamide's effect on skin. But things moved very rapidly from there! Since then, a flurry of studies have come out showcasing just how versatile this molecule is. Studies have shown that niacinamide boosts ceramide production, reduces hyperpigmentation, and even reduces the severity of acne lesions.

Niacinamide Around the World

How popular is niacinamide? It seems to be the molecule that everyone is on board with. It's one of the staple ingredients in many K-beauty brands. It was one of the top search terms in China in 2018. In fact, it's so popular that you'll find it in drugstore brands like CeraVe, clinical brands like PCA Skin, and even prestige brands like SK-II and La Prarie. So, yeah, niacinamide is pretty stinkin' popular, which also means it's everywhere. Just check those ingredient lists to make sure you don't end up with too much overlap.

at topical concentrations of 2–4%. Not surprisingly, other studies show this active as an effective acne treatment that can even help reduce the appearance of pores. How niacinamide achieves this is not well understood. Sebum production is a complex process—there are some recent studies out there that try to understand where niacinamide may come into play in a ten-page-long pathway. We'll eagerly await the results and give you guys the Cliff Notes later.

It can help tackle unwanted pigmentation: Niacinamide has been clinically demonstrated to effectively treat hyperpigmentation at topical concentrations as low as 2%. What makes niacinamide's brightening prowess special is the biological pathway that it targets. Unlike most brightening superstars—like hydroquinone, kojic acid, and vitamin C, which work as tyrosinase inhibitors (ahem—more on this on pages 205–213)—this guy actually works downstream from the tyrosinase step to deter pigment transfer. It targets melanosomes and prevents these cells from transferring pigment. This means that instead of competing for the same position, niacinamide works alongside other brightening superstars, boosting the efficacy of your products.

So, lots of good things, right? It gets better. Of all the actives, this ingredient has the best temperament from a Chemist's perspective, meaning it's easy to work with in almost any formula—creams, gels, serums, toners, you name it. This means pretty much everyone can reap some benefits from having a bit of niacinamide in their lives. So, let's figure out the optimal way to utilize this baby in your routine!

 Niacinamide really seems to be the ultimate multitasker. It keeps your skin barrier healthy, helps keep oily skin in check, and manages unwanted pigmentation. What can't it do?

How much niacinamide do I need?: Niacinamide has been shown to be effective at as low as 2%, with many of the tests done at 4%. So, a good target would be a 4–5% niacinamide product for most people, with 2% being a good starting point for those with sensitive skin. Because niacinamide happens to play nice with most actives, many brands can easily toss this into any of their products. Do remember to skim your products' ingredient lists to get a sense of how much niacinamide is already in your existing routine.

A recent trend has been serums touting 10% niacinamide and more. That's actually overkill! Even vanilla, gentle ingredients like niacinamide can be irritating at unnecessarily high levels. Please proceed with caution should you decide to give one of these highly concentrated products a go.

DEBUNKED:
The Great Debate About Niacinamide + Vitamin C

If you go down the Google rabbit hole on all things niacinamide, you'll stumble upon the great debate of whether or not you should be combining niacinamide and vitamin C.

- The issue: Vitamin C (ascorbic acid) and niacinamide are both clear when solubilized in water. But when the two are combined, the molecules interact and form a yellow liquid.

- And that's ... bad? In the "Don't Mix" Camp, there are thoughts that this interaction causes both ingredients to lose their efficacy or irritate skin. However, there's a study that actually shows that the combo still has skin benefits.

- So, it's a nonissue? For the most part, we wouldn't stress too much about this combination at all! In fact, having both in your routine is a powerful one-two punch to tackle hyperpigmentation.

INGREDIENT CORNER: Vitamin B5

When we talk about the ABCs of skincare, the *B* that we typically refer to is vitamin B3, otherwise called niacinamide. But, there's another vitamin B in the skincare world that deserves some attention.

You might recall in our moisturizer chapter (see pages 46–69), we mentioned pro-vitamin B5, panthenol, which happens to be a great moisturizing ingredient with multiple hydrating benefits.

Humectant: Helps skin keep hold of water.

Soothing ingredient: Calms dry, irritated skin.

The one catch about panthenol is that it's a really thick goo! So, expect creams with a lot of panthenol to lean on the heavier, more nourishing side. The good thing is that even just a little panthenol can be helpful, so consider this a little bonus to help skin stay calm and moisturized. All in all, despite sharing the same *B* letter, panthenol and niacinamide work quite differently, but they both work well with others to do great things for your skin barrier.

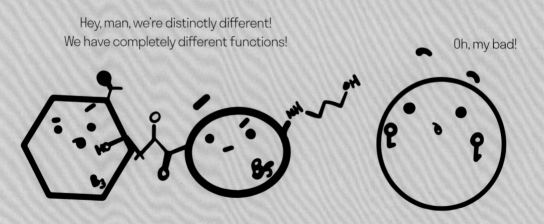

SKINCARE DECODED • 166

HOW DO I CHOOSE A NIACINAMIDE PRODUCT

To recap, niacinamide really does seem to do it all. For dry, compromised skin, niacinamide can help support a healthy skin barrier. For oily skin types, it can assist with oil control. It can even help tackle unwanted pigmentation and uneven skin tone. All of that at just about a 2–5% concentration. So, how do we get this ingredient into our routine?

Check In with Your Existing Routine

Start by reading the labels on your existing products. You might be surprised to find that you already have a few products in your arsenal that contain this ingredient. The key to incorporating niacinamide into your routine is to make sure you don't end up overlayering too many niacinamide products. Because it's such a prolific ingredient, you'd be surprised how easily you can overdo it without even realizing you've been layering eight niacinamide-boosted products.

Aim to Get Niacinamide in Your Moisturizer

One of niacinamide's greatest strengths is that it can play nice with everyone. Instead of dedicating a whole product step to a niacinamide serum, consider a chemist-favorite move and leave niacinamide to your moisturizer. Plus, you end up saving yourself a step in your routine. Time is money, people!

> All right—I definitely have to have this ingredient, but first let me check in with my routine real quick!

SHOP LIKE A CHEMIST

Always give the ingredient list a quick scan. If the product doesn't list a percentage, look for niacinamide in the top seven or so ingredients for an effective level. Other than that, niacinamide is really the most easygoing of the actives we cover here. There really isn't much else to consider formula-wise, other than personal preference. It's a truly chill fellow.

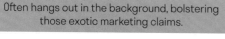

Often hangs out in the background, bolstering those exotic marketing claims.

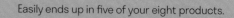

Easily ends up in five of your eight products.

Now that it's gaining fame, you'll see products that have gone overboard with the niacinamide.

Let's not forget its best trait: pairing really well with the other Big Four members!

CHEAT SHEET
Niacinamide Summarized

Chemists' Guidelines:

- Niacinamide, a form of vitamin B3, is the jack-of-all-trades in skin benefits. Data shows that it helps strengthen the skin barrier, reduces oiliness, brightens skin, and reduces pore size.

- Because this ingredient plays nicely in most formulas, it can get dumped in all sorts of skincare products, but you only need minimal amounts—a 2–5% concentration.

Recommended Starting Point by Skin Type:

Beginner

Sensitive   2–5% is the way to go for every skin type! Easy-peasy!

Oily

Chemists' Hacks and Pro Tips:

- More and more products are using 10% niacinamide concentrations and higher. While most people won't experience many issues at this concentration, just remember you don't have to hit those levels to see benefits, and there's no evidence that using five times more niacinamide gives you five times the benefit.

- Screen the products in your routine—you might be surprised where niacinamide hangs out.

PERSONAL TAKE

Gloria

Confession time? Niacinamide is possibly my least favorite of the Big Four. It really doesn't seem to do anything for me. When I cut out any of the other ingredients, I feel a difference. With niacinamide, I just never really miss it when it's not in my routine. However, many people swear by it, and niacinamide sure has the data to back it. Seriously, the number of studies niacinamide has almost solidifies its "literally Jesus" status—yet I don't notice a difference at all with or without it. This just highlights that everyone's skincare journey is truly personal!

Victoria

Gloria said what?! Well niacinamide works wonders for me! I've found that my skin handles high-level treatments much better with niacinamide around. But none of that serum stuff. I'm a lazy soul who needs as few steps as possible in my skincare routine, so I purposely look for moisturizers with niacinamide and save the treatment step for one of the other Big Four members!

NIACINAMIDE FAQs

1

Q: I found this cream that's 30% niacinamide. What are your thoughts?

A: There is such a thing as too much of a good thing in skincare. Most niacinamide studies use 2–5% of the ingredient. There's no reason to think 30% niacinamide is six times more effective as 5% niacinamide. You're more likely to irritate your skin with ultra-high concentrations than anything else.

2

Q: Are all niacinamide products alike?

A: Niacinamide is a very vanilla, stable ingredient. Unlike with vitamin C and retinol, you don't have to worry about its stability. And unlike acids, you don't have to think about pH too much. So, other than concentration, most niacinamide-based products are fairly similar.

3

Q: Is there a difference between the vitamin Bs used in skincare products?

A: The two vitamin Bs commonly found in skincare are niacinamide and panthenol, two completely different molecules. While panthenol doesn't fight pigmentation like niacinamide, it's a much better hydrating ingredient. We love the combination of the two!

4

Q: Will niacinamide overlighten my skin if I overuse it for hyperpigmentation?

A: Nope! It regulates pigment transport, but won't stop it.

5

Q: Can too much of it lead to excessive oil control?

A: Nope! This is the coolest thing about niacinamide: It actually improves barrier function, so it regulates sebum *and* improves hydration at the same time.

OTHER ACTIVES

The skincare industry is always chasing the next "miracle" ingredient. Outside of the Big Four, which we've just discussed, there are thousands more actives out there that somebody is (or no doubt will soon be) marketing as your must-have Fountain of Youth. And this is where it can get pretty confusing. These ingredients cover the full spectrum, from efficacious botanicals to absolute snake oils. You'll find anything from a garden full of botanicals to more exotic ingredients such as placenta! This category is both crowded and exhausting, but we'll go through it together and try to make sense of this zoo.

WHY OTHER ACTIVE INGREDIENTS

We just spent a ton of pages on the Big Four for their tried-and-true efficacy in treating a wide array of skin concerns. But once you start venturing outside of the Big Four, you've entered the Wild West of actives. Here, the language is more flowery and the results are murky. But before you walk away, we'll mention that there are still a number of great actives in this category that can work alone or alongside your core actives. So, we've put together a table of these actives that either are trendy or that we think are noteworthy, and we will go through how to pick these ingredients out and what to look out for.

A solid way to incorporate these ingredients is to pair them with a couple of the Big Four to tackle your skin concern. Just remember Chemists' Commandment #9: Patch-test with any new product.

DECODE That Label

WHAT TO LOOK FOR ON THE INGREDIENT LIST

Welp! All the guidelines we just discussed in the chapters on the Big Four active ingredient categories pretty much revolve around the importance of using the right concentration of the active ingredient for your skincare needs. This is relatively easy when we're talking about well-established, clinically proven molecules like niacinamide or L-ascorbic acid. But when we get into botanical extracts, all bets are off. The basic idea is that a botanical extract is a blend of helpful ingredients, diluted in some sort of solvent (most commonly water). The blends will vary depending on desired effect, and the quality of those extracts themselves may vary depending on a number of factors. It's part of a chemist's job to pick out the best one. But what this means is a *lot* of confusion when you try to figure out if a brand (1) uses high-quality extracts to begin with and (2) includes them at the right levels. A general rule of thumb is that you'll want to see your key extracts in the first half of the ingredient list.

ACTIVE SNAPSHOT

	Anti-aging	Soothing	Antioxidant	Hydrating	Acne
Bee Venom	-				+
Caffeine	-				
Centella Asiatica		+			
CoQ10			+		
Epidermal Growth Factor	-				
Galactomyces Ferment Filtrate	-				
German Chamomile		+			
Green Tea		+			+
Lactobacillus Ferment	-				
Lipoic Acid			+		
Low Molecular-Weight Hyaluronic Acid	+			+	
Manuka Honey		-		+	
Milk Thistle		+	+		
Peptides*	+	+			
Propolis Extract				+	
Royal Jelly				+	
Superoxide Dismutase			+		
Snail Mucin	-	-		+	
Turmeric		+	-		

* It depends on which peptide. Not all peptides are created equal.

Key

NOT Claimed	+	GOOD Potential
POPULAR Claim	-	Needs More Testing

Centella Asiatica
(also known as cica or tiger grass)

Skin benefits: This fantastic soothing ingredient has even shown some efficacy in wound healing.

Product tips: It's a *hot* ingredient that you'll find playing a star role in many soothing serums and creams. The confusing thing, though, is that there are a lot of ingredients that all derive from the same plant. These range from a generic extract to isolated compounds such as madecassoside. We recommend choosing a serum or moisturizer with one of the ingredients listed below in the top half of the ingredient list.

What to look for in the ingredient list:
- **Choose this:** Madecassoside, asiaticoside, asiatic acid, madecassic acid
- **Not that:** *Centella asiatica* extract, *centella asiatica* leaf water

German Chamomile

Fun fact! Most botanical ingredients you see on this list are Team Water ingredients. However, German chamomile is a unique, dark-blue essential oil with a light, pleasant scent.

Skin benefits: The plant's main benefit is soothing skin. Interestingly, a few studies have shown that it can also reduce hyperpigmentation. A great support ingredient all around!

Product tips: Definitely go with a product that uses bisabolol instead of the plant's essential-oil form. Bisabolol may not get much of the glamorous limelight enjoyed by other trendy competitors, but it's a solid ingredient with a long history of proven efficacy! Because it's oil-based, you'll see it in many soothing moisturizers.

What to look for in the ingredient list:
- **Choose this:** Bisabolol
- **Not that:** *Chamomilla recutita* (matricaria) flower oil

Turmeric

Skin benefits: Turmeric, with its Ayurvedic, traditional-medicine background, seems to offer benefits in *everything*—it's said to be an antioxidant, soothing, anti-pigmentation, anti-acne, antibacterial—you name it. From more modern clinical studies, this is best as a soother, with the potential to curb inflammatory acne. (Take the acne bit with a grain of salt.)

What to look for in the ingredient list: Isolated compounds and extracts offer higher concentration and less irritation than the essential oil form.
- **Choose this:** Tetrahydrodiferuloylmethane, tetrahydrocurcumin diacetate, *Curcuma longa* root extract
- **Not that:** Turmeric essential oil

Milk Thistle (Silymarin)

Skin benefits: This is one of the most interesting botanicals out there! It has great data for topical skin benefits, and, in fact, one study showed that a 1.4% silymarin product applied topically can reduce hyperpigmentation on par with pigment fighter hydroquinone. Its even bolstered by a study that shows strong anti-aging benefits targeting the glycation pathway.
- **Choose this:** *Silybum marianum* extract, *silybum marianum* fruit extract
- **Not that:** *Silybum marianum* seed oil

A special note on soothing ingredients

We see soothing ingredients as an important counterpart to aggressive, fast-acting active ingredients such as retinol and glycolic acid. Overly irritated skin leads to a cycle of inflammation and dryness that is counterproductive to long-term skincare. So, having a good blend of soothing ingredients in your routine is an important part of every skin type's daily upkeep. Other than botanical extracts, we also recommend scoping out allantoin and colloidal oatmeal. These are actual OTC ingredients recognized by the FDA for their skin-protectant properties. All of these, including *Centella asiatica*, German chamomile, and milk thistle are proven soothers that also come with some bonus side benefits like antioxidants or skin brightening.

Green Tea

Skin benefits: You can probably find claims about the wonders of green tea in a gazillion products out there. Fact is, in clinical testing, green tea has shown soothing benefits and, interestingly, even the ability to tame angry, inflamed acne. Many products tout green tea as an anti-aging ingredient with all sorts of amazing antioxidant benefits. But beware, this really hasn't been proven beyond the petri-dish phase.

What to look for in the ingredient list:

This one's tough! There are lots of different sources of green tea extract that all have the exact same ingredient name. It can be a bit of a guessing game as to who is using the most potent form. For a safer bet:

- **Choose this:** Epigallocatechin gallate (EGCG), *Camellia senensis* leaf extract in the top half of the ingredient list
- **Not that:** *Camellia senensis* leaf water

Where does it come from?

Plants: German chamomile, turmeric, *Centella asiatica*
Microbes (bacteria, yeast, fungi): Lactobacillus, galactomyces ferment filtrate, ceramides
Animals: Snail mucin, lanolin, bee products

BEE INGREDIENT FAMILY

Manuka honey: Popular for its wound-healing properties, but that popularity has really been exploited for its marketing fame in skincare. So, let's leave this one in the pantry.

Propolis: Popularized by K-beauty. Most of the studies are in petri-dish, animal, or open-wound settings—we're not clear what this actually does for your skin.

Royal jelly: Used for its hydrating properties.

Bee venom: Often positioned as an anti-aging ingredient. Interestingly, there are a few clinical studies on using bee venom to treat acne. Ultimately, it's still a pretty mysterious ingredient where a lot of the data is proprietary.

DECODE That Label

Antioxidant Corner

Antioxidants are a trendy skincare category that often gets inspiration from food trends. This results in skincare products that include all sorts of plant ingredients, from blueberries to exotic botanicals. Remember that our go-to recommendation in this category is L-ascorbic acid,, but there are a few notable alternatives! We recommend using them in your morning routine, under your sunscreen.

Resveratrol: This is the antioxidant found in grapes. A lot of brands will tout it as an antioxidant, but don't use enough to have any real effect. Look for more derm or science-backed brands that have a concentration of around 0.5%.

Lipoic acid: We've seen a good study showing that 5% lipoic acid can effectively reduce signs of photoaging. Given the high use level, we wouldn't recommend choosing a product that has lipoic acid lower than the fifth position in an ingredient list.

CoQ10: This is the abbreviation for "coenzyme Q10," also known as ubiquinone. It is a natural and integral part of your cell function and is an important part of your skin's natural defense system against outside aggressors. Topically, you ideally need about 0.5% CoQ10 in combination with vitamin E.

Superoxide dismutase: Another enzyme that is part of your body's natural antioxidant defense system. This is a very popular ingredient, but we are on the fence. It has wound-healing and soothing properties, but that data was collected on compromised skin, and it's not entirely clear what its cosmetic benefits would be. Ultimately, this wouldn't be our first skincare antioxidant of choice.

ALLEGED MIRACLES

Many of the ingredients on these two pages probably sound familiar to you. They are all gems in the eyes of skincare marketing execs, but from a chemist's point of view, their efficacy varies widely.

Galactomyces Ferment Filtrate

Galacta-who??? This is actually the ingredient behind SK-II's famed and proprietary Pitera. You can find it in a lot of K-beauty products as well.

Skin benefits: Another confusing, catch-all ingredient. Based on available clinical evidence, its main benefit is to smooth skin texture by regulating sebum production.

What to look for on the label: This *needs* to be within the first three or four ingredients on the list to have enough concentration to make an impact!

Snail Mucin

Skin benefits: Supposedly, snail slime cures acne, banishes wrinkles, improves skin texture, hydrates, and … leads to world peace, maybe? Sadly, there aren't really any scientific studies to back up these claims. It's fine as a hydrating addition … but not a must-have or must-try by any means.

What to look for on the label: *Mehhhh*—if you're looking for more visible results, look elsewhere. We highly recommend looking up how they extract the mucin, though; it's a fascinating video all around!

Lactobacillus Ferment Lysate

This and related lactobacillus ingredients are some of *the* most common ingredients used to claim "microbiome support." It's best to think of this as a great skin barrier support ingredient that will help your skin better retain moisture in the long run.

Caffeine

An oldie . . . but not really a goodie. For an ingredient that's super prevalent in skincare, there's a shocking lack of data showing that it does much. In other words, take claims about caffeine eye cream with a grain of salt. And take caffeine-containing anticellulite lotions with a tablespoon of salt!

Peptides

A super-difficult topic to write about from a consumer standpoint. There's a lot of interesting data on their wrinkle-fighting capabilities. But often, the data is proprietary and incredibly difficult to decipher for the layperson. We recommend sticking with the more established trade names in the peptides world, such as Matrixyl and Haloxyl.

Epidermal Growth Factor (aka "the Penis Facial")

EGF started trending in 2019—partly due to the facial treatment's popularity with Hollywood types, and partly because of that really unfortunate nickname. (Fine—we know you are curious.) According to the interwebs, the secret ingredient is stem cells that are no longer required by recently circumcised babies. Hey, you asked!) Give this trend another solid five years to see if any real, reliable science backs it up before jumping on the bandwagon.

Low Molecular-Weight Hyaluronic Acid

This popular hydrating ingredient is a very large molecule that sits on top of the skin and grabs hold of water. There are anti-aging products out there that use much smaller versions of the common hyaluronic acid and position it as more of an anti-aging material. The idea is that smaller hyaluronic acid penetrates more deeply and effectively plumps skin. There's actually quite a bit of data that backs this up. The catch? This can actually be irritating for some people. So, definitely patch-test before using!

IN PRACTICE

The best way to think of ingredients in this section is that these are the best supporting cast members. They may not be as good as the Big Four at treating targeted problems, but they can offer secondary benefits that may help soothe skin or enhance the efficacy of your core actives. When checking out the ingredient list, here are a few things to keep in mind:

1. Molecule > extract > water: Generally speaking, this is how we would rank the "usefulness" of an ingredient. So, for example, if you want a soothing, anti-inflammatory botanical, we recommend a product with bisabolol (isolated active molecule of German chamomile extract) over chamomile extract. Similarly, for the soothing benefits of *Centella asiatica*, choose madecassoside, or even *Centella asiatica* extract, rather than *Centella asiatica* leaf water.

2. Concentration!: Concentration still matters when it comes to extracts! Unfortunately, this is a very abused category. Generally speaking, you want to choose products with the claimed extracts at least in the top half of the ingredient list. Give the brand bonus points for simply sharing how much they're using! Yay, transparency!

3. Take extracts "rich in xyz" with a tablespoon of salt!: We did a few math exercises in the previous sections to showcase why this claim is one of our biggest pet peeves. "Rich in . . . " is one of those claims that sounds nice but, when inspected with a more critical eye, doesn't mean much, really. "Rich in vitamin C" is just not the same as "5% L-ascorbic acid."

4. The cocktail effect: In many cases, you'll get a synergistic benefit from combining actives. (So, for example, you might think that when combining 2 actives, you'll get a $1 + 1 = 2$ result, but a synergistic effect can mean that $1 + 1 > 2$). This is why we love brands that have done clinical tests on their proprietary blends!

5. Be skeptical of the newest trends: There's a lot of exciting new research in areas such as the skin's microbiome. However, products are usually three steps ahead of the science. If you see a brand-new trend pop up, we generally suggest waiting for gen 2 or gen 3, when the data becomes more robust!

Put Skincare Trends to the Sniff Test

The skincare industry is always chasing that latest, greatest, sexiest trend. We get it! It is part of the fun of shopping for skincare. But let's put the trends up to the chemists' sniff test and see which ones are potentially helpful and which ones are concentrated BS:

Should I really worry about it?: A big chunk of new trends comes from scare tactics. Something like, "Did you know that this thing is actually crushing your collagen and causing your face to age prematurely? Well, we've got just the miracle for you." Some damage claims are based in science; others are greatly exaggerated.

Blue Light	Infrared	Pollutants
Actual damage overblown. Ingredient science lacking. Hard for us to care a lot about this one.	Damage done in the long run is pretty well established. Antioxidants such as resveratrol are proposed solutions here.	Actually has good clinical studies showing the damage they can cause. Sunscreens and antioxidants are the way to go!

Is this hyped from other industries?: If something becomes a trend elsewhere, not long after, someone will try to put it in a face cream. This isn't inherently bad, but just know that this usually means the first few generations of product have practically zero scientific data on just what it does for your skin.

CBD	Adaptogens	Superfoods
At the time the first edition was written, there was a lot of hype around CBD application in skincare. Since then, wheels have thoroughly fallen of the hype train with little more data to share. Whomp whomp.	We have yet to see anything compelling in this area yet, although there's definitely interest out there.	From acai berries to kale, this category's all about antioxidants. We still recommend going with the tested ones as your core antioxidant strategy, but these don't hurt as support ingredients.

Biological sci-fi: This trend loves using buzzy science words! From DNA repair to mitochondria protection, this is a confusing arena where the frontier of skincare science meets absurd marketing claims. This category is definitely hard to crack as a consumer, so always look for a good clinical study.

Growth Factors	Stem Cells	Microbiome
Really new—almost too new to really say whether they work. This is where the penis facial stemmed from. Look for a good aesthetician that you trust if you want to try this facial.	It's actually a pretty old trend, and there's quite a variety of stem cell ingredients. It's generally lukewarm in market reception and results. We recommend looking at product clinical tests to get an idea of actual skin benefits.	Really promising new realm of skincare science, but the product execution isn't quite there yet. Keep an eye on this in the next five years for a breakthrough.

Rubbish Alert!

Raise a highly skeptical eyebrow if you read any of the following claims.
Anything that involves cellular improvement or aggressive, overnight results should be considered highly suspicious.

- "Instant rejuvenation"
- "Wipe years away in just one week"
- "Restructure DNA damage"
- "Activate energy production"
- "Instant skin bleach"
- "Powers your mitochondria"
- "24k gold for your 24k skin"
- "Targets your fibroblasts"

PERSONAL TAKE

Gloria

I have a love/hate relationship with this section. "Botanical that works" is such a hot topic, especially with the natural, clean movements in skincare. But the reality is, most botanicals have incredibly sparse data to begin with. It also bleeds over from food trends and other health trends. Ultimately, what we wanted to showcase in this chapter is a list of ingredients that may be worth trying . . . and some that have shockingly little data despite their popularity. Most ingredients here are good supplement ingredients to boost your overall routine. However, we consider these to be "sprinkles" on top of a more heavy-hitting foundation of proven ingredients.

I pretty much approach this category like I approach sheet masks: It's just all in good fun. New, trendy actives keep skincare fun and exciting, and I'll always be curious about this realm, but I definitely serve my curiosity with a strong side of skepticism. While I won't be replacing my staple actives with any of these, and I'll roll my eyes frequently at some of these claims, I'd probably still give some of them a try . . . at the right price.

Victoria

Section 3: BUILDING YOUR
ROUTINE

ROUTINES 101

So, now that we've covered the how, why, and key ingredients in skincare products, it's finally time to piece everything together. Before we jump into our guides on acne, hyperpigmentation, and anti-aging, let's go through some ground rules on building your routine. Let's have a little Routines 101!

For all lost souls, we recommend starting with the three fundamental pillars of skincare: cleansing, moisturizing, and sun protection. Remember that establishing a simple, consistent routine is more helpful than trying to jump into a 15-step one. Less is more, and consistency is key! If you're overwhelmed, here's a quick reminder of the key things to look for in each of these categories:

Cleansers (pages 30–45): You need to find your main squeeze, that go-to cleanser that gets you through most days (minus taking off any serious long-wear makeup.) This cleanser should strike a balance between gentleness and cleansing power, leaving skin feeling clean but not too tight or dry.

Moisturizers (pages 46–69): Remember the three parts that go into a moisturizer: humectants, emollients, and occlusives. Everyone's perfect moisturizer lies in the balance of these three. The goal is to find a moisturizer that doesn't sit too heavy on your skin, but instead keeps skin healthy and hydrated. It also shouldn't be so light that you need to reapply it every two hours.

Sunscreen (pages 70–103): The most important thing to remember when it comes to sunscreen is that texture is king. Most people apply far less than the recommended amount of sunscreen, often because they don't like the way it feels. To make sure you have proper protection from premature photoaging, choose a texture you won't mind slathering all over and will reapply often.

Once you've established these three pillars, you'll have the foundation you need to start building in actives and tweaking products based on your skin needs—and, more importantly, you'll have a solid routine to fall back on as your home base. That way, whenever your skin suddenly decides to have a fussy moment, you'll know what worked. So, if skin is angry, simply return to home base!

Oily Skin

Oily skin can be a challenge, and there's a temptation to overcleanse and under-moisturize. Here's how to hit the right balance.

1. **Gel cleanser:** One of the most common things we see with oily skin is people feeling the need to wash more than twice a day, or washing with a cleanser that has a squeaky finish that leaves your skin bone dry. What you may not realize is that overcleansing can land your skin in a world of trouble, further throwing off your hydration needs. If this sounds familiar, try a more gentle, less drying cleanser for two weeks, then assess how your skin responds. It's likely your skin might initially feel oily by early afternoon. Stick with it for a couple weeks to allow your skin to adjust. A good surfactant starting point is cocoamidopropyl betaine for less chance of leaving a residual film.

2. **Gel cream or 100% water-based gel moisturizer:** Although you may feel like you don't need any additional moisture, oily skin doesn't equate to moisturized skin. If anything, it points to an imbalance and you may not realize your skin is actually dehydrated. You're going for all the humectants here—gel creams with minimal lightweight oils or silicones. In fact, silicones provide a lighter finish and can leave skin a little matte. For incredibly oily skin, or in humid climates, a 100% water–based hydrating gel can cover your hydration needs.

Many people with oily skin try using mattifying moisturizers and sunscreens. These products typically contain a lot of oil-absorbing powders. While the powders can keep skin shine-free for a fancy dinner, these ingredients do not offer real long-term benefits.

3. **EU or Asia chemical sunscreen:** Although our constant reminders to use sunscreen may feel like nails on a chalkboard, it's a must! Definitely look to chemical sunscreens, which tend to have a lighter, more oily-skin-friendly texture. Asian sunscreens also formulate for more humid climates, so you can find some beautifully light sunscreens there!

Dry Skin

The name of the game is layering!

1. **Cleansing oil or balm:** One perk for dry skin types is that there are several cleansing formats you can explore. Gentle surfactant systems, oil cleansers, and cleansing balms are all good starting places that provide a gentle cleanse without exacerbating your dry skin. A bonus is that cleansing oils and balms can be pretty great at taking off long-wear makeup, too.

2. **Moisturizers: Creams, oils, balms—bring on the layers!** With dry skin, it's important to have not just your water-based hydrators but also all the oils and balms to keep all that moisture in the skin. That means you'll become a layering master. The rule of thumb is always to start with your water-based products like hydrating essences and serums, then finish with a more occlusives-heavy moisturizer or cream. We also recommend keeping a balm on hand to tend to those pesky dry patches that flare up.

3. **Mineral and alcohol-free chemical sunscreen:** Beware of "dry-touch" or "matte" sunscreens. These tend to contain ingredients such as denatured alcohol and oil-absorbing powders that'll make your skin feel parched by midday. Do give the ingredient list a quick scan and make sure you don't find alcohol too high up on the list. Mineral sunscreens are also worth a browse—these can have a heavier texture. Just try to patch-test on skin, since mineral formulas are not all created equal and can vary greatly in finish.

Special "Skin-arios"

Combination Skin

It can be such a pain picking products for combination skin! You have to worry about flaky patches and preventing midday shine at the same time. Here are a few tips to manage your mixed bag of skin needs:

1. **Gentle gel or cream cleanser:** The right cleanser prioritizes being gentle enough not to upset your dry skin zones, rather than emphasizing a more powerful cleanser for your oily skin zone. If you feel the need for a boost in cleansing power, consider using a cleansing brush.

2. **Lighter-occlusive moisturizer:** Many "oil-free" moisturizers do in fact use oil-based ingredients such as a lighter, occlusive-like dimethicone. These gel cream formulas are a great fit as a lighter moisturizer staple. Consider adding a balm to spot-treat any dry patch areas. Avoid any moisturizers claiming to be "oil absorbing," since they can exacerbate your dry patches.

3. **Sun protection:** This is a tough category that comes with a lukewarm answer. Find a light-enough texture you'll apply frequently. For those prone to dry patches, avoid drying sunscreens with lots of alcohol or powders.

Key ingredients to consider

Niacinamide: Our friend niacinamide not only increases your skin's hydration long term but also regulates sebum production. It's a match made in heaven for those with combination skin!

Mandelic acid: Don't worry about having to settle for a lukewarm AHA situation that aggravates dry patches while not quite tackling your oily zones. Mandelic acid hits the right balance—gentle on dry skin and yet still great for oily combination skin.

What If My Skin's Feeling Sensitive?

Most of us will experience skin irritation at some point in our lives. It can pop up from traveling, changes in weather, after undergoing a heavy-duty treatment, or just plain ol' aging. Typical signs of irritation include lingering stinging, pain, redness, changes in skin texture, and breakouts. With skin irritation, it's important to prioritize tending to your compromised barrier before turning to any aggressive active ingredients such as chemical exfoliants or retinoids.

Soothing ingredients to consider

Skin protectants: Colloidal oatmeal and allantoin have such a good track record, they are classified by the FDA as "skin-protectant" ingredients. They are effective ingredients that can help ease irritation in the long run.

Botanical soothers: There are a lot of plant-based soothers to consider in skincare! Some chemist favorites to consider include madecassoside and bisabolol.

Is it time to call a doctor?

Keratinocytes play an important part in your skin barrier, keeping water in and bad guys out. Once pathogens start interacting with the top layers of the epidermis, your keratinocytes are also responsible for sounding the alarm, sending signal proteins for leukocytes (white blood cells) to eliminate the pathogen.

As the first line of defense for our body, keratinocytes are extremely important for the immune system, but when cells get overreactive, they can cause complications. Overreactive keratinocytes bring about chronic inflammation, which can show up as eczema, rosacea, and even acne. Even if chronic inflammation goes unnoticed, it can still impact skin texture and morphology. Boo!

The point of this nerd-out session was to inform you that when skin inflammation feels like a chronic battle, it's best to bring in the big guns and seek out a dermatologist. You might need expert help in identifying your triggers and obtaining prescription care.

TROUBLESHOOT YOUR ROUTINE

Sometimes your skin can do a complete 180—it gets really dry for no apparent reason, or you move to a balmy new city, and, suddenly, your face is way too oily. Here, we'll give you some quick scenarios and our chemist take on how we would solve these annoying moments. The takeaway here is for you to get to a place where you're confident in your ability to adjust and edit your routine as needed. You are the expert on your skin!

Cleanser		
During or After Washing	*After Washing*	*After Washing*
Scenario One *"It burns!"*	**Scenario Two** *"My skin feels dry."*	**Scenario Three** *"It doesn't feel like it actually cleans." or "My skin feels heavy soon after."*
It's possible your skin just doesn't like this particular surfactant system. Look for a low-pH cleanser with a different surfactant.	Your current cleanser's cleansing power might be too much. Seek out a more gentle surfactant base. Consider a cleansing oil or balm.	You essentially want to up the cleansing power. This requires some trial and error, but coco betaine is a safe starting point. You can also consider a cleansing device to up your cleansing game.

Moisturizer		
Scenario One *"It's not moisturizing enough."*	**Scenario Two** *"It's too heavy, and I'm oily by afternoon."*	**Scenario Three** *"It feels moisturizing at first, but I feel dry by afternoon."*
Add a layer, any layer! A humectant serum or face oil, depending on your preference. A balm or salve is your best friend if dry patches develop.	Lighten it up. Scan your ingredient list to get a sense of the ratio of humectant to oils to occlusives. Go for a more humectant-forward product such as a watery serum or gel.	Your ratio might be off. You're aiming for the same hydrating finish but may need more of the emollients/occlusives to keep water within the skin.

Sunscreen		
Scenario One *"It's still too greasy."*	**Scenario Two** *"It's irritating."*	**Scenario Three** *"My skin gets congested, which makes me not want to reapply."*
Browse the Asia and EU sunscreens for lighter textures.	Switch systems. If you're using a chemical sunscreen, try minerals to see how your skin responds.	Try a lower SPF product. It's important to find a product that you're willing to reapply to get proper protection.

Adjusting Your Routine to the Seasons

The perfect routine may only be perfect as it adapts to changes in the weather. Here are some basic guidelines:

 Cold and dry: With less sun and a greater potential for dry skin, the theme here is nourishing. Additionally, with less sun, this makes for an ideal scenario to target your pigmentation. Consider the following products for your cold-and-dry routine: cleansing oils, balms, pigmentation actives. Consider dialing down your AHAs.

 Hot and dry: The complication here is that you need both lightweight and hydrating. Watery humectants won't cut it here. Consider the following products for your hot-and-dry routines: light cream cleansers, lightweight face oils, and AOX serums.

 Hot and humid: It's all about the water stuff. Actives treatments are a free-for-all here, so it's a great time to bring out all those anti-aging serums and essences. Consider the following products: light chemical sunscreens, gel cleansers, gel cream moisturizers, and anti-aging serums and essences. Use micellar waters and toners to freshen up midday.

ACNE

For many of us, acne is the reason why anyone starts paying attention to their skin in the first place. This taboo subject plagues so many of us—teenage blunder years, anyone? Annoyingly, acne can even be a continual skin concern well into your forties. Moderate to severe acne will require you to team up with a dermatologist to truly tackle breakouts. However, there are thousands of acne topicals available in drugstores to help you try to zap the buggers. So, let's try to make sense of this massive category to filter out the acne duds and establish some ground rules for your acne-fighting routine.

ACNE BIOLOGY: A GENERAL PICTURE

Forget what you've heard about how bacteria growth is the root cause of acne. Historically, many people believed that acne is caused by an overgrowth of bacteria called *Propionibacterium acnes*—now renamed *Cutibacterium acnes (C. acnes)*. This misconception inevitably resulted in regimens packed full of oral and topical antibiotics. In the end, *C. acnes* turned out to be only a piece of the puzzle, and these treatments alone are not going to cut it. So, let's dive deeper and take a look at the sebaceous gland, the main site of acne growth.

Sebaceous glands are crucial to healthy skin because they secrete anti-microbial lipids, upregulate antioxidants, maintain pro- and anti-inflammatory responses, and secrete sebum. However, they also happen to possess just the right scenario for acne growth for two specific reasons:

1. **Annoying hormones:** Sebum secretion is promoted by an increase in androgens—which is why acne is a classic trademark of puberty. This class of hormones is responsible for triggering sebaceous-gland growth and sebum production, prime material for acne breakouts. They also are responsible for hyperproliferation (aka excessive cell growth), which can cause buildup within the follicle. Double trouble!

2. **Triggered inflammation:** Sebaceous glands do have pro-inflammatory lipids, which help as a defense mechanism but can also make an acne situation worse. In fact, it can even cause acne without any sort of *C. acnes* overgrowth.

All of this means that *C. acnes* is more the result of an unhappy sebaceous gland than the cause. The takeaway here is, acne is truly a complex condition that isn't easily solved by washing your face four times a day or dousing breakouts in super-concentrated topicals. We now understand that acne can stem from various triggers such as genetic disposition, hormonal imbalance, and stress, with links to androgens and pro-inflammatory lipids. With this multitude of factors, acne has forever been a frustrating and lengthy process for treatment, requiring a lot of patience, diligence, and constant trial and error.

YOUR SAVIOR, THE DERMATOLOGIST

As we've mentioned, acne breakouts are not easily solved with just drugstore topicals. Ultimately, if breakouts become unmanageable, a derm will be your one true partner who can guide you through prescription treatments. Let's run through a quick checklist of when to consult a derm!

If you're experiencing any of the following scenarios, it's time to seek a derm!

- Drugstore topicals are causing significant irritation.
- Your longtime topicals aren't cutting it anymore; your breakouts show no improvement.
- There's a drastic change for the worse in your acne breakouts.
- The number of acne lesions is increasing.
- Acne lesions are red or painful to the touch.
- Lesions are forming deeper in the skin.

What if you feel like you're not getting results with a derm, or the regimen is getting stale? While we *do* want to remind you of Chemists' Commandment #4 (All good things take time), if you've been at it for six months with no improvement, we *also* want to remind you that there is nothing wrong with a second opinion.

Do I Have Fungal Acne?

If you venture down the interwebs rabbit hole for information on acne, you may eventually find yourself wondering whether you have fungal acne (aka Malassezia folliculitis). First, know that the fungus responsible for this infection is a type of yeast called *Malassezia*, and, as gross as it sounds, Malassezia is naturally found as part of our skin's microbiome. While an imbalance of this yeast can lead to a skin condition that often looks like acne, what Google doesn't tell you is that it's very hard to actually diagnose without the help of a derm. So, the takeaway here is that if you've reached this conclusion trying to solve your stubborn acne breakouts yourself, step away from the laptop and get yourself into a derm office. Don't start reconstructing your routine to be fungal acne–safe—there's no clear consensus on what ingredients are good for this skin condition.

BUILDING YOUR ACNE ROUTINE

When you're at the drugstore, you may be overwhelmed by the plethora of acne topicals available over the counter. Before breaking down the ingredients, claims, and formulations, we should go over some points that may seem contradictory to things we've said elsewhere in the book:

1. **Treating acne typically requires a cocktail of ingredients:** You will most likely need multiple ingredients to team up and tackle your breakouts. Like other active treatments, introduce your acne-fighting products one at a time. Take the time to see how your skin reacts before trying to add in the next actives treatment.

2. **Refrain from exorcising your face of all oil:** A lot of acne products are all about being "oil-free," often touting oil reduction or oil control. But all skin needs oil! Even if you have acne. The lesson here is, try not to get crazy with all of these oil-control products. Acne-prone skin is delicate and easily compromised, so don't wash your face too many times a day, use clay masks too often, or constantly reapply oil-absorbing powders.

3. **Try not to pick at lesions:** We get it—sometimes you just can't help but pop a pimple. If you absolutely can't help yourself, please wash your hands, keep it sterile, or just leave it to the professionals—aka aestheticians!

4. **Cleansers with actives are helpful here:** For acne, actives in cleansers have been shown to be a helpful mode of treatment. That means you have one more slot in your routine to add in another acne-targeting active, since cocktails of multiple actives are important here for managing breakouts. Consider using a salicylic acid or benzoyl peroxide face wash to maximize your routine.

Routine Builder: Acne

Here, we'll lay out a framework for how you can build your acne routine. Ultimately, you'll want to hit a diverse group of goals:

1. **Healthy cell turnover to expedite acne lesion healing**
2. **Antimicrobial to minimize *C. acnes* overgrowth**
3. **Soothers to keep skin calm while dealing with acne and strong topicals**
4. **Post-acne fighters to manage residual pigmentation**

If you're using prescription topicals, that's even better! Not only do you have tailored topicals prescribed by an expert, you've also got an expert on hand to run through your routine with you. In fact, bring in your products with you when you visit the derm so they can give you the green light on your routine with their prescribed treatment.

Day Routine

In your day routine, you have two possible steps for acne ingredients: cleanser and treatment step. For daytime, try to keep it simple with ingredients like salicylic acid and benzoyl peroxide. One product that acne-prone skin types purposely like to forget is sunscreen. We get it—the texture, shine, and oiliness makes you want to do a hard pass, but sunscreen is really important for post-acne healing. Too much sun can lead to those spots darkening. On top of that, a lot of these acne-fighting ingredients make you more sensitive to the sun, so let's sunscreen, people! To keep things light, you can consider forgoing your moisturizer step for sunscreen if you're struggling with oiliness.

Salicylic Acid Cleanser → BPO Treatment → Moisturizer (If Needed) → Sun Protection

Night Routine

Adapalene or other retinoids are great night-treatment additions. These tend to be more aggressive and can cause irritation or dryness. A helpful step to troubleshoot is to start your routine with a soothing serum.

Salicylic Acid or AHA Cleanser → Soothing Serum (Optional) → 0.1% Adapalene Treatment → Moisturizer

A note on post-acne care

It's easy to want to panic when looking at your pigmentation after acne has come and gone. Good thing the pigmentation section is up next! And for those wondering if it's an acne scar, any change in skin topography is considered acne scarring, and a derm will be better able to asses your type of scar and which in-office procedure is best. Most scars won't be solved with just topicals alone, which makes a great case for us to stop picking at those pimples.

Pro tip: A weekly clay mask can serve as a dual-purpose step here. Not only can it help absorb excess oils, but removing the excess layer of oils can also improve the penetration of your topicals.

INGREDIENT HIGHLIGHT
SALICYLIC ACID

This is one of those ingredients that has been around for centuries as a headache and fever remedy. Historians have found evidence it was used throughout ancient history from China to Egypt to Rome and even prehistoric North America. Today, it's a hot skincare ingredient for more than acne because of its ability to exfoliate at the pore level while still being an anti-inflammatory. As more and more brands catch on, this means that it can often end up in all sorts of products. Much like niacinamide, salicylic acid is potentially found in cleansers, treatments, oils, and even moisturizers. Just make sure you give the ingredient list a quick glance to make sure you're not overlayering salicylic acid. There is such a thing as too much of a good thing!

ACNE FIGHTERS

Here are some of the common acne-fighting ingredients you can find on the market without a prescription. They can treat mild acne alone or work in tandem with your prescription care.

Heavy Hitters			
	Concentration	**Description**	**Chemist Notes**
Adapalene	0.10%	A new OTC retinoid with promising data against acne but significantly diminished side effects compared to tretinoin	Suggestions of a synergistic effect with BPO. Consider pairing these two in your routine.
Benzoyl Peroxide	2.5–10%	OTC ingredient that can be a great spot treatment. Get ready for bleached pillowcases!	With the BPO offering, it's easy to want to reach for the max-strength 10% benzoyl peroxide option. But there are some studies that suggest more doesn't actually equal better. 5% is a good starting point as a stand-alone topical for sparse breakouts and general acne maintenance.
Azelaic Acid	20%	Convenient that this ingredient also has great data on combating PIH	20% azelaic acid is pretty hard to find in stores. Instead, you'll often find a 10% option. At 10%, consider this more as an acne-maintenance ingredient than an effective breakout treatment.

Helpful Secondary Hitters			
	Concentration	**Description**	**Chemist Notes**
Sodium Ascorbyl Phosphate	5%	Vitamin C derivative often found in serum form	Can serve as a dual-purpose pigmentation and acne treatment for those acne-prone-skin types looking to incorporate some vitamin C into their routine.
Tea Tree Oil	Max 10%	An essential oil that surprisingly has some results in minimizing breakouts	In pure form, this essential oil can be very irritating for skin, so be sure to properly dilute it. Can suffice as a spot treatment. 5% tea tree oil was found to perform on par with 5% benzoyl peroxide but required longer treatment time.

Exfoliants			
	Concentration	**Description**	**Chemist Notes**
Salicylic Acid	0.5–2%	OTC regulated. This is the ingredient that gets dumped into every product.	Typically better as a support product to another topical. Try starting at the lower end of the spectrum. High-level 2% salicylic acid topicals positioned for acne can often be very drying, which can further aggravate inflamed acne. Decode the ingredient list to find a less drying version.
Alpha Hydroxy Acids (AHAs)	5–15% daily	AHAs include glycolic, lactic, and mandelic acids. Aid in increasing cell turnover, which can be helpful for speeding up acne recovery.	Typically better as a support product to another topical.

Pro Tip: Looking for a soothing product to boost your acne-fighting routine? Consider green tea extract. Like many plant extracts, this ingredient has antioxidant and soothing properties. More interestingly, this has been specifically found in tests to reduce the severity of acne. A great ingredient to add!

The Jury Is Still Out

With acne being such a chronic, tumultuous, spontaneous condition, it definitely comes with what feels like a lot of voodoo magic. So, the list below includes a few topical ingredients and diet restrictions that have been rumored to help but actually haven't been properly studied or shown significant evidence to work. (We'd also like to remind you of Commandment #10: Skincare is personal. If skipping that glass of milk seems to be working for you, stay the course!)

- Sulfur
- Willow bark extract
- Witch hazel
- Not eating dairy
- Not eating sugar
- Not eating chocolate
- Egg whites
- Toothpaste

PIGMENTATION

Brown spots, sun spots, melasma, uneven skin tone—whatever your pigment woes may be, we feel your pain. Pigmentation happens to be one of the most persistent and difficult skin concerns to tackle. It typically takes a whole village of products and a lot of diligence to see significant results. But hard work pays off, so let's build your routine to tackle this stubborn skin concern.

PIGMENTATION BIOLOGY: A GENERAL PICTURE

Before getting into the complex strategy of fighting uneven skin pigmentation, let's run through why uneven pigmentation even happens in the first place. The process is actually pretty straightforward, so we'll speed-walk through the biology so you'll have some context behind our strategy. Let's start with some key players.

Meet Mr. Melanocyte. Melanocytes are skin cells that live in the deepest part of your epidermis and are, in a nutshell, your skin's one-stop-shop for pigment. Through a process called melanogenesis, each melanocyte produces, packages, and delivers melanin to the skin's upper layers, where it becomes visible pigmentation. Ta-da!

Tyrosi-what? Tyrosinase! One of the most important enzymes in this process, tyrosinase determines how fast melanin is produced—and when it comes to melanogenesis, "how fast" is crucial. Uneven skin tone, melasma, and hyperpigmentation are all linked to overly enthusiastic melanin production.

External Triggers: The melanin-production process can be affected by a number of external factors that can cause a group of melanocyte cells to amp up production, leading to uneven skin tone and unwanted dark spots.

 Excessive UV exposure: When it comes to hyperpigmentation, the sun is enemy Numero Uno! It's the leading cause of unwanted pigmentation, triggering uneven, excess production of melanin.

 Inflammation: Irritation and inflammation are key triggers of hyperpigmentation. In fact, PIH (post-inflammatory hyperpigmentation) can be an unwanted side effect from chemical peels or laser treatments.

 Hormones: Our hormones can influence melanogenesis, which is why some women can exhibit melasma during or after pregnancy.

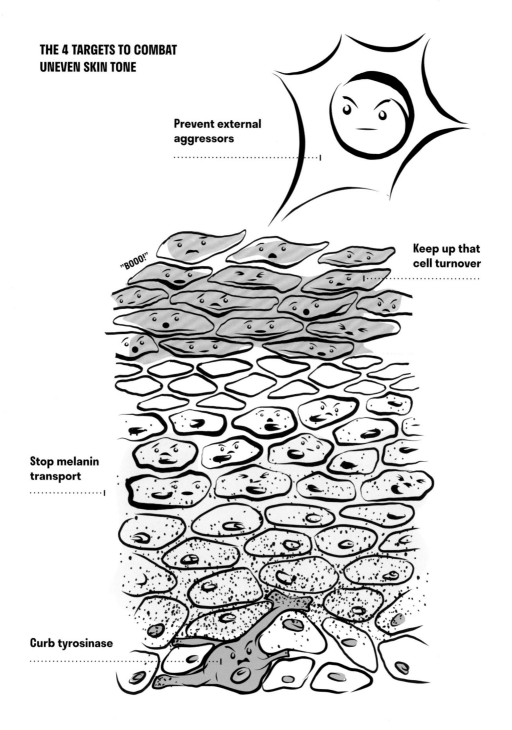

THE 4 TARGETS TO COMBAT
UNEVEN SKIN TONE

Prevent external
aggressors

Keep up that
cell turnover

Stop melanin
transport

Curb tyrosinase

BATTLE AGAINST HYPERPIGMENTATION

Pigmentation issues are so frustrating, in part because there's no quick fix. The good news is that a truly comprehensive, diligent skincare routine can effectively address the dark patches that can result from sun damage, aging, acne scarring, and other factors. A well-rounded and effective hyperpigmentation-fighting routine involves every step and every factor that might affect melanogenesis. Here are the steps to success in this battle:

1. **Prevent external aggressors.** The best offense is a good defense.
2. **Tell your tyrosinase to take a chill pill.** Slow down pigment production.
3. **Slow melanin transport to the upper layers.** Melanin doesn't become visible until it makes it to the upper layers. So, if no one can see it, then it doesn't exist.
4. **Make sure your cell turnover is in tip-top shape!** Chemical exfoliants shed your old, dull skin faster, helping to fade dark spots.

INGREDIENT HIGHLIGHT
Hydroquinone

While it's considered the gold-standard pigment-fighting ingredient, hydroquinone is also highly controversial in some circles, due in part to some scary misconceptions. So, let's do some hydroquinone myth busting!

I heard you need a prescription to use hydroquinone. That's right! Recently, the US has banned over-the-counter sales of hydroquinone. You'll now need a prescription to use this one.

Does hydroquinone really cause skin cancer? False! There isn't any evidence that hydroquinone causes cancer, although it isn't an ingredient you should use continually without supervision. It can be irritating and have side effects like the halo effect, in which a ring of lighter skin forms around the original dark spot.

The bottom line? Hydroquinone is still one of the gold-standard topicals used to treat many hyperpigmentation concerns, including melasma. In fact, it's better to team up with your dermatologist to monitor your progress, since it is best used under guidance.

ROUTINE BUILDER Pigmentation

Pigmentation is incredibly science-dense, with multiple facets to tackle. Even after understanding the why and how of hyperpigmentation and its associated skincare ingredients, building a routine to actually tackle the concern is a different story. If you're a lost soul, worry not! Consider our routine primer a Pinterest-able guide to building your pigmentation-product strategy.

Day Routine

Here, we have the foundation of your daytime routine. The key strategy is damage prevention, with the two key products being an antioxidant serum and a good daily sunscreen. L-ascorbic acid is the gold-standard antioxidant and tyrosinase inhibitor, making it a prime candidate for the daytime routine. For those of you who find it irritating, you can consider replacement ingredients such as silymarin, lipoid acid, resveratrol, or magnesium ascorbyl phosphate. You can use any sunscreen you like, just as long as you use it daily! Not only is sun protection key to preventing damage that spurs the cycle of hyperpigmentation, but many pigment-fighting actives can also make your skin more sensitive to sunlight.

Cleanser → L-Ascorbic Acid (Vit C) Serum → Moisturizer → Sun Protection

Level Up!: We get it, sometimes you just need more!!! Remember that AOX and sunscreen are the priority, so don't forget those steps as you layer in additional pigmentation-fighting ingredients. If you see any pilling or layering, try a different treatment—don't abandon your basics. Consider adding another dose of milder pigment-fighting ingredients, such as azelaic acid or botanicals like licorice root extract, or finding a moisturizer with niacinamide.

Cleanser → L-Ascorbic Acid (Vit C) Serum → Azealic Acid (optional) → Moisturizer → Sun Protection

Night Routine

Below is an example of a well-rounded night routine. This is really where you ramp up the actives to correct the unwanted hyperpigmentation!

Cleanser → Exfoliate (Glycolic Acid) → Tyrosinase-Inhibiting Treatments → Moisturizer (Consider Soothers and Niacinamide)

1. **Exfoliate:** Glycolic acid is really the gold standard here. This step doesn't necessarily have to be daily if your night routine is starting to feel overbooked.

2. **Tyrosinase inhibiting:** Nightime is a great time to target any hyperactive tyrosinase. Keep in mind that many highly effective ingredients can also be a bit irritating; this list ranks pigment-fighting actives by general irritation potential. However, many pigmentation-fighting serums contain a blend of these ingredients. Use this list as a loose guide, and always patch test!

   - **Mild:** Licorice root, Uva ursa (bearberry), peony extract

   - **Medium:** Arbutin, kojic acid, azelaic acid, tranexamic acid, resorcinol, ascorbyl glucoside

   - **Heavy-duty:** Hydroquinone

3. **Moisturize:** It's great to find a niacinamide-containing moisturizer. Niacinamide targets the melanin-transport step instead of blocking tyrosinase, which means it complements your serum perfectly.

Retinol-Centric Routine: For those who also want to loop in wrinkle prevention, retinol can be a great staple to build your nighttime routine around. Remember that retinol will require some acclimation, so it's best to pair it with a soothing serum. Turn to pages 175–178 for soothing ingredients to look out for during your next shopping trip.

Cleanser → Soother → Retinol → Niacinamide Containing Moisturizer

Meet Your Core Pigment Fighters!

There are a ton of pigment-fighting active ingredients out there. They all work to regulate or slow down pigment production one way or another. One of the most common targets for these ingredients is our friend tyrosinase. Since tyrosinase determines how fast melanin is created, telling it to take a chill pill is a very effective strategy.

Of course, tyrosinase isn't just going to take it without a fight. This is why most effective brightening treatments contain a blend of of ingredients. We classify these hyperpigmentation fighters into three general ingredient categories: isolated compounds, multitaskers, and botanical brighteners.

Isolated Compounds

These ingredients have been identified and tested as effective hyperpigmentation fighters. The key to these ingredients is not using them at such a high percentage that they irritate your skin. Many products come with a blend of active ingredients; we recommend choosing one that contains at least one of these key compounds.

	Chemist Notes	Target Concentration
Hydroquinone	Considered the gold-standard pigment-fighting active ingredient. See the Ingredient Highlight on page 207 for some hydroquinone know-how.	2% to 4%
Arbutin	Chemically speaking, a close relative of hydroquinone.	2% and up
Tranexamic Acid	Though relatively new to the US market, it's been popular in Asia for years. Fun fact! It's been administered as both oral tablets and injections to fight melasma.	2% to 5%
Kojic Acid	This is a fairly unstable compound. If you notice your product going from white to brown to black, please throw it out.	1% to 2%
Ellagic Acid	Oof—this is a difficult ingredient to work with. It's quite rare; you'll see it more in extract form, such as pomegranate extract.	1% or in the top half of the ingredient list
Resorcinol	This can be a whole family of ingredients with similar chemical structures. It's also used in chemical peels.	Often part of a proprietary blend

Multitaskers

Who doesn't love an ingredient that can solve more than one of your skin concerns? These ingredients are great additions to your other brightening active ingredients, depending on your skin needs.

	Chemist Notes	Target Concentration
Azelaic Acid	A very popular ingredient in recent years! In addition to data on pigment fighting, there's data on this ingredient for treating acne.	10–20%
L-Ascorbic Acid	We love it because it fights hyperpigmentation in more than one way, preventing damage as an antioxidant and slowing down tyrosinase productivity.	5–20%
Retinoids	Need we say more about this skincare superstar? Check out our chapter on Retinoids (starting on page 130) to find out which retinoid is right for you, and how to shop for it.	See page 138

Botanical Brighteners

"Natural" has been a hot category in skincare for the past couple of decades. The thing is, not all botanical extracts have been through rigorous testing—or much testing at all, really. Another difficulty with extracts is that it's hard to assess whether a brand has used an effective dose of active ingredients. For example, bearberry (*Uva ursi*) extract is a natural source of arbutin. But just how much arbutin is in the extract or how much extract is in the product is impossible to tell from the packaging alone. Sadly, this lets less responsible brands get away with using a sprinkle o' fairy dust's worth of the good stuff. Here's a list of our chemist-preferred botanicals that have a decent amount of data backing up their skin benefits. Think of these as great sidekicks for your isolated compounds. However, we wouldn't rely on these alone to tackle pigmentation issues:

- Milk thistle extract
- Licorice root extract
- Mulberry extract
- Bearberry extract
- White peony extract
- Emblica extract

Support Ingredients

Ultimately, unwanted hyperpigmentation is too stubborn, too complicated for just one good treatment to solve. It really takes being mindful through your entire routine for the most effective, efficient strategy. Other than your main hyperpigmentation treatment, there are three other areas of support you should consider: damage prevention, blocking melanin transport to the upper layers, and chemical exfoliation.

ANGLE ONE Prevent Damage

The best treatment for pigmentation is prevention, my friends! In the biology section (pages 204–207), we talked about how pigmentation is stimulated by outside factors like UV rays and inflammation. So, naturally, an integral part of your pigmentation care is to get ahead of these pesky unwanted spots before they evolve into full-fledged hyperpigmentation. Here are your product helpers and what to look for:

	Sunscreen	Antioxidants	Soothing Ingredients
Why It's Important	By now, you should realize how important it is to have good sun-protection habits to avoid pigment woes. Without sunscreen, a routine full of hyperpigmentation actives will be pretty useless.	These are the perfect products to pair with your sunscreen for some extra oomph to battle free radical generation. L-AA is our favorite here. It's an antioxidant *and* a tyrosinase inhibitor. Win-win!	Inflammation is a big source of excessive pigmentation, so an easy prevention technique is to do what we can to keep skin calm and happy. There are a lot of great botanicals in this category, but they are not all created equal.
Ingredients to Look For	For more information on how to choose a good daily sunscreen, check back to pages 84–85.	Ascorbic acid (vitamin C), tocopherol (vitamin E), lipoic acid, resveratrol.	Bisabolol, *Centella asiatica* (madecassoside, asiaticoside), allantoin, cucumber extract, etc.

ANGLE TWO Hold the Melanin

Of course, if you bombard your tyrosinase with all these actives to slow it down, at some point you're going to saturate this method. So, to help your tyrosinase fighters be more effective, you can target unwanted pigmentation by just, well, not sending it up. Melanin is made in the deepest layer of the epidermis, but this melanin doesn't translate to visible pigment until it gets delivered upward in a process called melanin transfer.

	The Big Picture	Concentration
Niacinamide	What can help slow down the delivery process? Our good old friend niacinamide. A great addition to your pigment-fighting skincare routine, it increases efficacy without taking anything away from your other brightening products.	About 2–5%. Nowadays, many products will have much higher niacinamide concentration. Be sure to proceed with caution to not overwhelm and irritate your skin with such high levels.

ANGLE THREEE Unveil New Skin

Your skin is constantly turning over, generating new skin. So, by using chemical exfoliants like AHAs, you help cycle through any unwanted pigment anomalies as soon as possible. A high-level chemical-exfoliant peel, or even a professional-strength peel, can help reveal brighter, more even skin tone faster.

Beware of going too aggressive too fast. Especially for those with darker skin, post-inflammatory hyperpigmentation from overaggressive peels is a real concern—not to mention that it defeats the whole purpose of getting a peel anyway. So, make sure you're acclimating your skin to such actives, and beware of anyone who tries to sell you on something overly aggressive right off the bat.

	The Big Picture	Concentration
AHAs	Glycolic acid is the gold-standard AHA when it comes to battling hyperpigmentation. However, it may not be the right molecule for everyone. Read more about other options in the chapter starting on page 114.	Approximately 5–10%

ANTI-AGING

As of today, there is no miracle, holy-grail face cream existing on this planet that reverses time. Now that we've cleared that up, let's talk about the dreaded wrinkles. There are actually a lot of solid actives that can help prevent wrinkle formation and minimize wrinkle severity. Adding a few wrinkle-fighting actives on top of a solid skincare routine (cleanser, moisturizer, sunscreen) will have you covered, and you can sleep peacefully, knowing your preventative, wrinkle-fighting routine is helping you age gracefully.

BIOLOGY OF AGING: A GENERAL PICTURE

There are two main culprits of aging: the sun and good ol' Father Time. Intrinsic aging from, well, just living is called chronoaging, while aging caused by external factors (mainly UV rays) is called photoaging. Clinically speaking, chronoaging is associated with thinning skin and loss of elasticity. On the other hand, photoaging is characterized by deeper wrinkles and sagging. Our faces are naturally exposed to the elements on a regular basis, so chronoaging and photoaging work in tandem and lead to changes such as fine lines, wrinkles, sagging, chronic dry skin, and unwanted hyperpigmentation.

In the dermis, you'll find crucial proteins responsible for holding everything up. Degradation in these proteins is at the crux of skin aging. Refer to pages 132–132 for more context.

Collagen: This protein, with its sturdy triple-helix coil structure, is *the* crucial support protein in your dermis.

Elastin: This appropriately named protein is responsible for your skin's elastic "bounce-back."

These proteins are the prime victims of aging pathways such as free radical damage, glycation caused by excess sugar, elastosis (elastin buildup), and the like. Over time, your collagen breaks down, modifications and cross-links build up, and an excess of elastin crowds your dermis. This internal structural damage translates to all the visible signs of aging, like wrinkles and sagging.

// Yikes! I'm not sure what all these words mean . . . but take my money and save my collagen! //

THE TRUTH
ABOUT ANTI-AGING

We know what leads to skin's intrinsic aging process, but can skincare really tackle this complex phenomenon happening deeper in the skin? It's complicated. Although collagen levels are key, measuring collagen deep in the dermis is invasive and complicated. Instead, most skincare studies assess different mechanical, functional, and visual characteristics of skin to capture overall skin quality.

Anti-Aging Evidence

One annoying thing about the anti-aging world is the high volume of overhyped products that come with a creative faux-science story. Think magical-unicorn-tears-in-a-bottle sort of lingo. There is one way to sniff out the gems from the sea of doodoo, and that's by looking at the evidence. The best brands will conduct studies and even take measures to track and prove the benefits of their products. So, let's learn all the ways you can measure a wrinkle!

Cutometer: This nifty little instrument is literally a fancy suction cup. It pulls skin up, then releases it. By measuring how far the skin can be pulled and how fast it goes back to its normal position, scientists can determine how "firm" and "elastic" it is. Pretty neat, huh?

Visia: This is essentially a fancy, well-calibrated photo booth. It can take a way-too-high-resolution picture of your face with different lights. The downside is that it highlights everything your phone's photo filters aim to hide, but it does allow scientists to calculate some pretty cool stats like wrinkle count, discoloration severity, and even sun damage.

Expert Panel: These are people trained to critically assess your skin conditions visually. While expert panels are not as precise as actual instruments, you would be surprised how consistent they are. Great to have on your team to evaluate products!

A well-designed clinical study that includes measurements like these is the ideal snapshot for understanding how an anti-aging product will perform. Sadly, there are very few products that have been rated using these tools. But good thing you have us! Let's build that anti-aging routine and take a more critical look at alleged anti-aging miracle ingredients.

Routine Strategy: Making the Big Four Work for Your Skin

Whether your goal is aging prevention or you're seeking active correction, your core skincare strategy should be to make the skincare Big Four (introduced on page 112) work for your skin type. Simply put, the Big Four just have the most scientific data showing that they work! Let's go through a few "skin-arios" and build a couple of chemist-recommended routines to start you on the right foot.

INGREDIENT HIGHLIGHT
Peptides

Other than the Big Four, one of the most common ingredients found in anti-aging products is peptides. As we mentioned in the treatments section, peptides encompass a vast category of ingredients that can be difficult to navigate. Most ingredients are highly proprietary, and the only data available are held by peptide manufacturers. So, let's go over some of the most common peptides you can find on the market:

Matrixyl 3000: You can find this on your ingredient list as palmitoyl tripeptide-1 and palmitoyl tetrapeptide-7, or even called out directly on the packaging. This is one of the most well-known peptides out there, created by Sederma, which has done multiple clinicals to showcase that Matrixyl can visibly reduce the appearance of wrinkles.

Haloxyl: This is another Sederma classic! Unlike Matrixyl, it has been tested more specifically for eye-area concerns such as dark circles and firmness around the eye area. You can spot this on your ingredient list as hydroxysuccinimide, palmitoyl tripeptide-1, and palmitoyl tetrapeptide-7.

sh-Oligopeptide-1 : This is one of the main ingredients found in products claiming "growth factor" benefits. It can be found alone or in a blend with other peptides. There are promising signs with this ingredient. However, since it hasn't gone through as much testing, we recommend scoping products that use this with a blend of other good ingredients, such as proven antioxidants, for a more comprehensive anti-aging case.

Routines for Your Skin-ario

Skin-ario 1: Aging Prevention

Are you in your twenties or thirties and starting to think about keeping your skin in tip-top shape? Then aging prevention should be your top priority! Daily antioxidant and sun protection are the most important parts of the prevention strategy. Consider starting here:

Day Routine

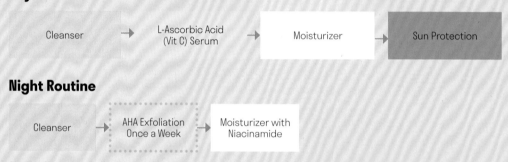

Cleanser → L-Ascorbic Acid (Vit C) Serum → Moisturizer → Sun Protection

Night Routine

Cleanser → AHA Exfoliation Once a Week → Moisturizer with Niacinamide

Skin-ario 2: Actively Fighting Fine Lines

If you're noticing fine lines starting to creep up and want to take a more active approach to anti-aging, consider incorporating a lower concentration of retinoid to your routine—for example, 0.3% retinol two to three times a week. If you're extra sensitive, consider alternatives such as Granactive Retinoid or bakuchiol.

Day Routine

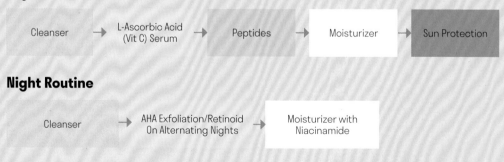

Cleanser → L-Ascorbic Acid (Vit C) Serum → Peptides → Moisturizer → Sun Protection

Night Routine

Cleanser → AHA Exfoliation/Retinoid On Alternating Nights → Moisturizer with Niacinamide

Anti-Aging Pro Tips

Pro Tip 1: It's still important to keep skin barrier function in mind when ramping up retinoids. Light irritation and shedding is expected during acclimation, but your skin shouldn't be in a constant state of red, itchy irritation.

Pro Tip 2: Aging skin tends to start skewing dryer. Consider scoping a more occlusive moisturizer by looking for ingredients such as petrolatum and shea butter higher up on the ingredient list as a night cream.

Pro Tip 3: Claims like "45% improvement in appearance of wrinkles after 8 weeks" are better than consumer perception claims like "93% of participants agree that skin looks lifted." Stories with mystical, proprietary ingredients with no claims are most likely not going to get you the right results.

Pro Tip 4: Vitamin C is the gold-standard antioxidant, but it can help to include a blend of different antioxidants. Refer to our antioxidant corner on page 153 for some of our favorite antioxidants! The good thing is, many antioxidant-packed products on the market come in blends, so you don't have to worry about layering different products!

Skin-ario 3: Leveling Up with an In-Office Procedure

Ultimately, in-office procedures will get you much more dramatic results in a shorter time than at-home treatments. If you decide to opt in for procedures, your post-procedural skincare can help it work that much better.

Recovery phase: If your skin is very angry post-procedure, keep your routine simple and focus on moisturization and sun protection. Soothing ingredients like allantoin and *Centella asiatica* extract can also help.

Help the procedure go further: Fun fact! There are many studies that show home-use chemical peels can be a great treatment between office visits to help enhance the effects of your professional peel.

 Pro Tip: When it comes to preventing the signs of aging, your lifestyle matters! Study after study has shown that eating a high-sugar diet and smoking tobacco are both bad for your health and can negatively impact the appearance of skin.

IN-OFFICE VS. HOME-USE TREATMENTS

If you're ready to try out in-office procedures to treat wrinkles, it's time to consult a dermatologist or aesthetician. When you go in for your consultation, ask about these procedures and work with the derm to see which is right for your skin type and skin conditions:

- **Laser therapy**
- **Chemical Peel**
- **Dermabrasion/Dermaplaning**
- **Microneedling treatment**
- **Botox**

Laser Light Therapy: There are a LOT of in-office laser therapies to choose from that can target pretty much anything you can think of—from acne and hyperpigmentation to anti-aging concerns such as resurfacing and firming benefits. Home alternatives have gained a lot of popularity in recent years. The most common ones are red-light devices for anti-aging benefits and blue-light devices for treating acne.

The good news is that light therapy does have clinical data supporting its benefits, but these home devices will always be significantly weaker than their in-office counterparts. That means results with a home device will never be as significant as a series of in-office light therapy sessions. To get the most out of home LED devices, we recommend using the devices consistently. Devices that have independent clinicals get major brownie points from us!

Microneedle: Like many in-office procedures, microneedles are designed to "damage" the skin in a strategic manner to stimulate collagen production. This can be incredibly effective for improving the appearance of skin texture, fine lines and wrinkles, and even acne scars.

Thus, it's only natural that with the popularity of the in-office procedure, you can buy a range of home-use microneedles from almost any shopping channel for as little as $15. With the sheer number of testimonials showing dramatic before and after results, it's highly tempting to go down this avenue. However, most store-bought microneedles are *not* as fine as office-use needles, nor do they puncture as deeply. This means the wound created is neither refined nor as effective, which is not ideal, since that's the whole purpose of this method. Moreover, to get the most out of microneedling, having an expert assess your skin condition and microneedling correctly can make a huge difference in the results. Lastly, not all at-home microneedle devices are marketed correctly. All of these devices should be one-time-use products for sanitary reasons. With all of these negatives, home-use microneedles are a firm nope for us!

Chemical Peels: Chemical peels are treatments that use high levels of weak acids, such as glycolic acid, salicylic acid, or trichloroacetic acid, to penetrate deeply into skin. This helps reboot cell turnover and can also boost collagen production. In-office peels are much more aggressive than home versions. For example, a home-use glycolic acid peel can use 30% glycolic acid at a pH of ~3.5. An in-office glycolic peel can be as high as 70% with a pH of <1. This is definitely something that should be administered and monitored by a professional! We'd highly recommend staying away from home-use peels that claim to be at these professional concentration levels.

That said, this is a category where home-use versions are still quite effective. In fact, studies have shown that home peels can extend the effects of your in-office treatments. For more information, check out our chapter dedicated to AHAs (pages 116–127).

DECODE That Claim

"BETTER THAN BOTOX" SERUMS?

Botox is one of the most popular in-office treatments to help preserve a youthful appearance. From time to time, you can find topical treatments trying to ride the Botox wave and make bold claims such as "needle-free Botox!" or "put off Botox for ten more years!" As good as that sounds, the reality is, there is no topical that can mimic the effects of Botox. But good news! Anti-aging superstars like retinol have been tested to help enhance the effects of Botox. So definitely consider mastering your retinoid game!

THE EYE AREA

The biggest category next to anti-aging has to be eye creams and eye-area concerns. Maintaining a youthful eye area is one of the most coveted arenas of beauty, and those tiny little jars of eye cream come with lots and lots of promises! But, the challenge with eye creams is that treating these concerns (eye bags, crow's feet, dark circles, puffiness), is complex, not well studied, and probably the most difficult to see actual results. It doesn't help that an individual doesn't usually have just one of these concerns, but multiple issues. In this section, we'll go over some eye area biology to better understand the complexity of treating the eye area. We'll also analyze some of the most common ingredients associated with treating the eye-area and, finally, provide tips and tricks on how to navigate the eye-cream landscape and their waffling price tags.

WHY EYE-AREA BIOLOGY IS ONE BIG SPAGHETTI MONSTER

T The main eye area concerns that are targeted in skincare are crow's feet, dark circles, eye bags, and puffiness.

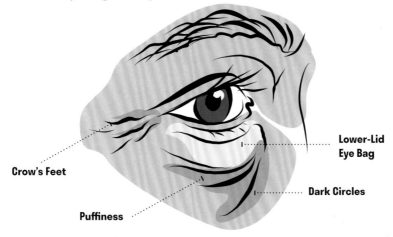

Crow's Feet

Puffiness

Lower-Lid Eye Bag

Dark Circles

Crow's Feet: These are wrinkles that form at the outside corner of the eyes. Wrinkles for the most part are a well-studied category in cosmetics. The good news is that some of the ingredients known to treat wrinkles also have decent outcomes in treating crow's feet. We recommend referring to the anti-aging and retinol sections (pages 130–145, 172–185) for guidance.

Dark Circles: This would probably rank second in terms of understanding and treatments. It's a complicated concern that can stem from volume loss in the tear trough area, skin laxity, translucent skin, excessive pigmentation, allergies, and/or vasculature changes. Of the research out there, we'll share the breakdown done by a research group in Taiwan that depicts the complexity of dark circles well. They have categorized dark circles into four types based on pigmentation and cause.

P-type: Responsible for brown hues. This is the same pigment that contributes to hyperpigmentation and melasma.

V-type: Responsible for blue, pink, and purple hues. It's common that this type of dark circle pigmentation also lends itself to undereye puffiness.

S-type: Responsible for shadow hues, typically caused by a change in the structure of the eye area that leads to a hollowing effect. It can also cause eye bags to develop.

M-type: Typically a combination of the three types listed above. Most of us who have dark circle concerns usually fall into this category.

Puffiness and eye bags: Interestingly, undereye puffiness and eye bags can be difficult to differentiate. The general theory is that eye puffiness is usually located in front of the fat pad and is caused by vascular leakage. This fluid buildup can also contribute to purple, pink, or blue pigment hues. On the other hand, eye bags usually result from a change in the structure of the fat pad itself as well as our bone structure. Annoyingly, it is also possible to have both concerns at the same time.

Ultimately, understanding the biology helps us untangle and structure the way to think about our eye-area concerns. However, the real challenge is how to treat these concerns. You've been warned, topical data is sparse at best.

COMMON EYE CREAM INGREDIENTS

Because there isn't a strong foundation of understanding around eye-area concerns, the research on treating these concerns is even more unorganized and sparse. Here is an update of ingredients that have been looked at in published research:

Vitamin K (phytonadione)

Area of Concern: Dark circles
Level of Evidence: Weak

This is a pretty popular active in research that gets mentioned a lot in dark circle care. However, there is such a lack of evidence that it's hard for us to gauge how effective this active really is. The only clinical we could really find was a study carried out in 2015 in which one group tested a 3% caffeine + 1% vitamin K on 11 subjects (read: tiny!) where they applied the treatment pad to one eye while they applied water to the other eye. After four weeks of use, they found that subjects saw a ~16% improvement from the baseline. To be honest, with how small the test group was, the relatively minimal improvement, and short test duration, we're not convinced.

Caffeine

Area of Concern: Dark circles and puffiness
Level of Evidence: Weak

For some reason, this ingredient continues to trend as a major eye-cream active, but we have never been able to figure out why. The theory on caffeine is that this ingredient causes vasoconstriction, which should reduce the amount of fluid (aka puffiness) around the eye area. Well, instead of reading our rant about why we think caffeine is useless, we'll let a Thai research group tell you their feels. In 2010, this group tested a 3% caffeine treatment on 34 subjects who went without a night of sleep. Their conclusion: "However, the overall efficacy of the selected caffeine gel in reducing puffy eyes was not significantly different from that of its gel base. It could be concluded that the cooling effect of the hydrophilic gels was the main parameter in reduction of eye puffiness rather than the vasoconstriction of caffeine."

Of the few studies we could find on caffeine, most were blends with other actives or had paper-thin data sets. We'll leave caffeine to rest here.

Vitamin C (ascorbic acid)

Area of Concern: Dark circles
Level of Evidence: Promising?

Probably the best paper of the bunch that captures how stubborn dark circles are was done by a group in India in 2016. This very small study of 16 melanin-rich subjects split them into 3 groups: Group 1 received monthly 20% glycolic acid peels (gasp!); Group 2 received monthly 15% lactic acid peels; and Group 3 applied vitamin C nightly. They found that 73% of subjects in Group 1, 57% of subjects in Group 2, and 27% of individuals Group 3 saw improvement. While we can't make hard conclusions from such a small study, it does give us a decent summary of how challenging this concern is even when using some heavy-hitting concentrations of these tried-and-true actives.

*Caution: We do not recommend using these levels of AHAs on your eye area at home. We only get one set of eyes, people.

Okay, we basically just gave you a list of pretty lackluster ingredients. Sadly, this is the state of ingredient research for eye creams. However, keep in mind that, since the biology is multifactored, it actually makes sense that no single ingredient is able to truly tackle eye concerns in a big way. It also means that you are now empowered with the knowledge not to fall for eye creams that sell you some mystical ingredient that promises to erase your eye-area problems. (We're looking at *you*, caffeine.)

Instead of hunting for single ingredients, we recommend looking at blends of actives and the potential clinical data that comes with them. This means we will have to heavily rely on brands and their product testing to give us a sense of performance. How do you know if the testing is up to snuff? We can give you some clues!

NAVIGATING THE EYE CREAM LANDSCAPE VIA TESTING

Because there isn't enough third-party research or solid active-ingredient knowledge to help organize the eye-cream landscape, you'll have to look at product marketing for clues on performance. Even a transparent ingredient list won't be able to tell you much in terms of efficacy. Instead, we look to the quality of an eye cream's claims testing, which we've categorized into three tiers:

Tier 1. The "Get Religious" Category

These eye creams come with zero testing—that means no consumer testing and no clinical testing. This really is just a Hail Mary. Just say a little prayer and hope it'll do something.

Tier 2. The "Mid" Eye Cream

These eye creams come with consumer perception studies. We consider this the B team of testing, simply because of its anecdotal nature. Consumer perception studies are self-assessment studies where subjects come in at specific time points and self-evaluate their experience and any improvements they may have noticed. To read more about consumer perception go to page 234.

> **NOTE:** Many eye creams that rely solely on consumer perception will still use the lingo "clinically tested." Read your data carefully!

Tier 3. The "Gold-Star" Eye Cream

This eye cream comes with actual measured clinical data! You may hear lingo such as "___% subjects saw significant improvement" or "study participants saw ___% improvement." Something to keep an eye out for (hee hee, get it?)

here is the length of the study. We'd give studies between six to twelve weeks a lot more weight than shorter studies. We see some tests as short as two weeks with "clinical improvement." This type of timeline definitely tickles our skeptometer. Even though we all want to see dramatic results ASAP, patience and consistency over *at least* four weeks are still key to seeing real improvements.

THREE COMMON EYE CREAM CLAIMS TO LOOK OUT FOR

When shopping in the land of eye creams, you'll realize that some can cost just as much as a face cream for half the volume. A lot of times the markup for eye creams doesn't seem quite justified, so how can we avoid overpaying for these little jars? We have found three common claims that can give clues to help you avoid duds and pick out an eye cream with a better chance of giving you the results you're looking for.

1. The Faux Clinical Claim: "100% showed a brighter under-eye area*"

Reading the asterisk is key! Upon reading the asterisk, you find that this was after one use. We call this a faux clinical claim because no eye cream will brighten eyes after one use unless the product uses optical powders (think makeup). Many eye creams claim these rapid results, but you'll often find they include soft-focus powders like boron nitride, silica, iron oxides, etc. that are used in makeup to give a blurring, lightening effect. These types of instant effects can be helpful, especially when paired with a concealer during the day, but we wouldn't call this a true clinical claim that will bring a long-term benefit.

NOTE: Don't ignore the (*) when reading testing data. The * tells you a lot about testing parameters and the length of the study. It can be very eye-opening (hah. Another eye pun).

2. The Not-So Clinical, Clinical Claim: "Up to 100% reduction in dark circles after 8 weeks"

A claim like 100% reduction sounds pretty convincing! But when you read more general wording like "up to," proceed with a healthy dose of skepticism. This could mean that in a ~30-subject study, maybe only one of those subjects saw a 100% reduction, while the rest of the group saw very little reduction. If you stumble on this type of claim, look to see if any other testing claims are shared to get a better idea of performance.

3. The Clinical Claim: "97% showed a reduction in puffiness after 8 weeks"

Hooray! We've hit a legit clinical claim! This claim tells us there was marked improvement via instrument or expert grader after eight weeks. To make this claim even better, we'd like to see statistical significance. The claim would then change to "97% showed a significant reduction in puffiness after 8 weeks." It's a one-word difference but makes it *significantly* more convincing. (This is the last pun ... maybe.)

FINAL THOUGHTS

On top of the complex biology, factors like genes, diet, health, and even bone structure also contribute to these eye-area concerns, and some of these are outside of our control. Thus, satisfaction with your eye cream really comes down to managing expectations. Will eye creams fade your dark circles, puffiness, and eye bags 100%? No. That degree of improvement will require an in-office procedure. But good eye-cream formulas can probably help lighten your dark circles somewhat and minimize wrinkles. Ultimately, we view eye creams as an enrichment product to your skincare routine rather than a must-have and advise you to consider a great concealer sidekick as well!

Eye Area Hydration!
There's nothing wrong with using your face moisturizer or serum for your eyes. Just use caution with those runny serums, and proceed with a lot of caution if the skincare formula contains AHAs or retinol.

SHOPPING GUIDE

This is supposed to be a practical field guide to skincare, so we definitely have to dissect the skincare shopping experience. In this last section of the book, we'll walk through our best tips on how to make a more educated purchase for skincare at all price points, how to shop based on your values, and even more ways to decode skincare product ingredient lists. If you've been wondering about things like "How do I find a better moisturizer for my skin?" or "How much should I actually be paying for a good moisturizer?"—read on!

DECODING INGREDIENT LISTS

In our journey of skincare science education, our feelings toward decoding a skincare ingredient list (IL) have definitely evolved a lot! Initially, Chemist Confessions got its start by helping people make sense of all the scary-sounding names they find on an ingredient list. But, now, we've realized that in some instances, decoding has gotten carried away. There are even a few instances where decoding created much more confusion than being helpful. Ultimately, ingredient lists are just another piece of the product research puzzle. They can give you clues to potential skin-reaction triggers, help you decide between products, and, ideally, save you some time and a little money.

Rules to Help You Decode

Order of Ingredients: Ingredient lists are (mostly) organized from highest concentration to lowest concentration. This is helpful to remember when you want to get an idea of how much active or moisturizing ingredient you might be getting.

The 1% mark and phenoxyethanol: The catch with the ingredients order is that brands are free to rearrange ingredients for those of less than 1%. This can actually get confusing, especially for lower-use level actives such as peptides and retinol. If included, the commonly used preservative, phenoxyethanol, can be a good indicator of the 1% mark, since it should be used at a maximum of 1%. This means the amount of any ingredient that comes after phenoxyethanol will be either at 1% or less. We should also mention that brands are becoming more strategic at obscuring the 1% mark, so this isn't always a surefire indicator.

Speaking of phenoxyethanol, know your preservatives: For those with sensitive skin, preservatives is a category of ingredients to keep an eye out for. If you're doing detective work on what may be your skin triggers, start by paying attention to similar preservatives in products that cause an allergic reaction. Some common preservatives in skincare include parabens, phenoxyethanol, ethyhexylglycerin, sodium benzoate, sorbic acid, chlorphenesin, and anisic acid.

When Decoding Ingredients Is Helpful

Treatments and Serums: Of all the skincare categories, decoding the IL for this group is probably the most helpful. Concentration always matters! You need enough of the active ingredients to get results, but you also might want to avoid too high of a concentration that could cause skin irritation. Additionally, this is a category where claims can be pretty misleading and you'll find the marketed ingredient maybe isn't what you thought you were getting. Try to keep tabs on the concentrations your skin needs for the Big Four (retinol, ascorbic acid, niacinamide, and chemical exfoliants).

When Decoding Ingredients is Kind of Helpful

Moisturizers: Decoding each ingredient in a moisturizer won't tell you much unless you're actively avoiding a very specific ingredient. However, a general sense of a moisturizer's humectant, emollient, and occlusive ingredients can give you an idea of what kind of hydration you're getting. For example, there are gel cream moisturizers that might not be suitable for dry skin types because they have a much heavier amount of humectants but very few emollients and no occlusives. You can also stumble on thick creams that seem to have a lot of butters with very few humectants. These imbalances could be the reason why you feel like your skin isn't completely happy with your hydration routine, and you'll be able to adjust accordingly. Refer to our moisturizer chapter (pages 46–69) for common ingredients in each of these categories.

When Decoding Ingredients is Not Helpful

Cleansers: As chemists, we can tell you that cleansers can be very complicated and nuanced formulation projects. Surfactants are used in unique blends with carefully chosen supporting ingredients to fine-tune the final cleanse experience. This makes troubleshooting cleanser irritation pretty difficult because not only are you not sure which surfactant in the blend is giving your skin trouble, but it can also be concentration-dependent. For example, SLS (sodium lauryl sulfate) is a demonized surfactant because it happens also to be used in clinical studies to intentionally cause irritation on the skin. However, you can create mild, highly effective cleansers with low levels of SLS. This is why decoding cleanser ingredients isn't necessarily helpful, since cleanser preferences are very personal.

Sunscreens: If you find yourself decoding a sunscreen label, aside from choosing between a mineral or chemical sunscreen, it's time to close out those tabs and stop the research. All sunscreens must go through FDA-regulated OTC testing, so trying to read into sunscreen labels won't give you much more information. As formulators, we can tell you that sunscreens are one of the most challenging formulas to make. Any ingredient in the formula can impact formula stability and SPF levels. So there's really no point in trying to hone in on any particular ingredient unless you're actively avoiding a specific irritant for you. Choose your sunscreen based on texture!

Decoding Testing Claims

Ingredient lists are only one arena of clues you can use to help choose your next skincare product. The information (or lack thereof) on labels can actually give you a couple of hints on whether or not the product is right for you. With a few tips and tricks, you can dissect marketing claims and decide for yourself whether or not the product really warrants the price tag and will truly help tackle your skin concern.

In the past couple of years, more and more products are making claims based clinical testing. But, sadly, we still see wildly aggressive claims based on confusing interpretations of tests. Let's go through the types of testing evidence you might see out in the wild and decipher what they *really* mean:

Before and After Pictures: Probably considered the wildest of the bunch. You can find anything from pictures taken with what seems like a flip phone to legitimate clinical photos. We find before and afters often show the best-case scenario but definitely would not give you a sense of what the average result would be. Be skeptical of before and afters with lighting differences, filters, and skin-tone differences. The ideal before and after would be a professional photo with standardized face positioning. Ultimately, if you see a great before and after, we recommend thinking of this as a promising start, and to keep looking for more data to come with it.

Consumer Perception: This is a type of clinical study where participants will use the product for a certain amount of time and then are required to fill out a self-assessment questionnaire about their product use. You can tell a marketing claim is a consumer perception study when they use lingo like "x% agree" or "x% saw an improvement." These claims typically use a humanized verb. We would say this is a mid-ranked claim. It gives you a more concrete idea of consumer experience compared to a typical customer review, but it's still hard to gauge how much of a skin benefit you'll get.

Clinical Studies: This is the ideal type of study that usually involves selecting a group of participants to trial the product and then return to the clinical site to have an expert grader or instruments measure results at specific time points. These claims are the closest we can get to understanding how a product on average would perform on skin. However, these results will never be as exciting as consumer perception results. You'll find claims of a ~38% wrinkle reduction (which is really great!), but it definitely sounds less exciting compared to "99% agree."

Unfortunately, when you're shopping around, you'll find that very few products come with actual clinical testing. There are a few possible explanations here:

- Probably the biggest reason is that clinical testing takes extra money and testing can be very expensive.
- The product format doesn't really warrant clinical testing unless you have free money lying around. Face cleansers come to mind.
- The brand doesn't feel that clinical testing is necessary to sell its product. This makes us sad.

This is why we give major brownie points to products that have been put through a clinical test. In fact, for any serums, treatments, and products that come with a cushy price tag, we really emphasize looking for any sort of testing.

The Skincare Shopping Commandments

Even with the few clues to product labeling, shopping for skincare can still feel like you're swimming in an open ocean. There are so many retailers, e-tailers, and opportunities to purchase skincare. Exciting! But you can easily become inundated with options and suddenly find yourself in a completely different product category. You were shopping for that hydration piece and now you ended up in ... toner land?! We've summarized all of the product knowledge we've just shared into a few simple rules to help you hunt for the next product.

1. DO read between the marketing claims! Clinical testing is the closest you can get to an objective result when it comes to marketing claims. It can give you a more accurate picture of what you can expect and how long you'll need to use the product to see results.

2. DO decode ingredient lists for serums and treatments.

3. DO NOT decode ingredient lists for cleansers and sunscreens.

4. DO keep track of key percentages for the Big Four.

5. Know your main active percentages, specifically, actives like retinol, vitamin C, salicylic acid, and acne topicals.

6. DO keep track of ingredients in your moisturizing routine.

7. DO quality check your product periodically. Did it turn lumpy? Did the product thin out? Is there a layer of oil seeping out? Don't let these signs slide!

8. DO manage expectations. Beware of any product that claims to "erase" skin concerns in a matter of days.

DECODE That Claim

One of the most important shopping criteria is deciphering product claims, and skincare products make a lot of them. Some are helpful, and many others ... not so much.

"Clean" Claims:

Organic: This claim isn't very meaningful. The organic claim is declared by the Department of Agriculture and not the FDA. For plant-based products, most are so refined that we're not too concerned with trace pesticides. Like, at all.

Natural: "Natural" is a popular claim on the market. However, it's important to remember that natural ingredients are not inherently safer or more effective than their synthetic counterparts.

Vegan: No animal-derived ingredients were used in the formula. Generally, okay, though there are a few ingredients that are animal-derived that can be very helpful to skin.

No-no lists: It can be confusing (and scary!) when you see every other brand come out with conflicting and lengthy "no-no" lists of "dangerous" ingredients. Know that most of these ingredient dangers are blown way out of proportion. It's best to focus on what ingredients work and don't work for your skin, rather than following any one brand's guideline.

"Free From" Claims:

Chemical-free: Everything, even water, is a "chemical." It really isn't as scary as some marketing lingo wants you to believe!

Gluten-free: Your skin doesn't absorb gluten-based ingredients like your stomach does. This is not an issue for those with celiac disease, since celiac disease involves your small intestine. Unless you're eating your skincare products—but then you've got other problems.

Preservative-free: Oh, boy, does this one rub us the wrong way. Your products *need* preservatives. Without proper preservation, your product can grow bacteria, yeast, fungi—all sorts of unfriendly microbes.

Paraben-free: Because of a single article that came out showing intact parabens in breast tissue, every country has extensively looked into parabens and their link to causing breast cancer. Despite all that work, no regulating body has shown any decisive evidence that parabens cause breast cancer. In fact, parabens happen to be one of the most gentle preservatives around because just a *little amount* can be very effective. The downside is that you'll rarely find a paraben product, and brands have had to find alternatives that aren't as all-around gentle as parabens.

Cruelty-free: As animal lovers ourselves, we think it's great that the market is embracing this claim. The good news is, nowadays, it is exceedingly rare to test finished products on animals. We do want to mention that many safety and toxicology ingredient tests are still tested on animals. But ending this practice is not as simple as just stopping the tests altogether right away—to ensure safety, it must be done carefully. Phasing out animal testing is a huge topic being actively worked on by scientists. It is still very much a work in progress.

Oil-free: This isn't a regulated claim, which means brands can decide how they define something to be "oil-free." It can span anywhere from having zero oil components whatsoever to "just as long as it doesn't have the word *oil* in it." For example, there are a few sunscreens that like to claim to be oil-free to make oily-skin types feel more at ease. The problem with this is chemical sunscreens are inherently oil-based. Just remember that the right oils can be great for even oily skin types.

"Medical" Claims:

Sensitive skin, hypoallergenic: There's actually no official guideline to make these statements, which means that products with these claims can or cannot aggravate sensitive skin types. Helpful, right? Ultimately, it will be up to you, the skincare user, to figure out which ingredients are problematic for your skin.

Non-comedogenic: Many consumers are under the impression this involves rabbit-ear testing. Even now, *non-comedogenic* doesn't always signify that an ingredient won't cause you to break out. Again, this is why it's important that you, the user, figure out which ingredients are problematic for your skin and decode those ingredient lists.

COSMETICS AND THE ENVIRONMENT

It's great to see that more and more of us are becoming value-conscious shoppers who want to do our part to take care of Mother Earth. But, as in other areas of environmental awareness, the beauty industry is challenging and comes with its own can of worms. So, without adding too many more pages to this book, let's briefly touch on some general facets of cosmetics and the environment.

Microplastics: Plastic exfoliating beads were probably a staple in a lot of people's skin-washing routines in the early 2000s. They're perfectly spherical, which means they offer a more gentle exfoliating experience than natural counterparts, such as apricot. Unfortunately, they end up in the bellies of fish and are generally a huge hazard to aquatic life.

The good news is that microplastics were such a big issue in the 2010s that laws are already in place that have banned the manufacturing and production of these beads in personal care. The bad news is, contamination from cosmetics is just a small part of the puzzle. Surprisingly, one of the worst offenders is not even from personal care—it's your actual clothes. Try your best to wear clothing made from natural materials. Avoid polyester, spandex, and Lycra, which typically take at least 20 years to degrade. Instead, look for natural/biodegradable fibers like cotton, wool, or rayon. And of course, when possible, we can all do our part by stocking our closets with well-made, long-lasting pieces instead of 574,289,571 cheap shirts that we toss out after two wears.

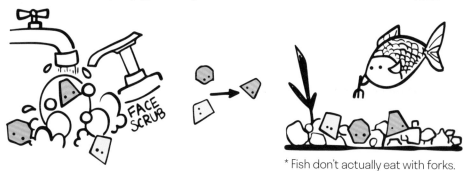

* Fish don't actually eat with forks.

Sunscreen: There has been a huge push to tackle the mammoth issue of climate change. A few studies have shown that certain chemical sunscreen filters *may* have an impact on coral reefs, specifically oxybenzone and octinoxate. However, the scopes of these studies have only just scratched the surface. There is still so much more that needs to be researched to truly understand the impact of sunscreen filters (including mineral filters) on our coral reefs. We should note that there already has been a huge shift to formulate away from these two sunscreen filters. Ultimately, it's important to keep in mind that sunscreen filters are just one very small component of coral reef preservation and we are continuing to monitor as more findings are shared.

Palm oil: When you think about palm oil, you can't help but also think about orangutans. Palm oil is a huge topic not only in skincare but in almost every industry. It's an affordable raw material with an absurd amount of applications in our modern daily lives. Think of your potato chips, shampoos, and definitely skincare. This huge demand has inevitably led to deforestation and habitat destruction.

The one piece of good news is that there are a few organizations hard at work promoting the sustainable harvesting of palm oil. The most notable organization is the Roundtable on Sustainable Palm Oil (RSPO.org). Check out their website for more insights on how to shop sustainable palm oil.

PLASTIC PACKAGING AND RECYCLING

Another area you might like to think about when shopping for skincare is package sustainability. The use of virgin plastic is growing at an alarming rate; however, we're still in the midst of figuring out how to reduce virgin-plastic consumption and improve our recycling management. For now, we'll simply cover what's feasible at this moment.

Recycling: If you think about skincare packaging, you might feel that these components are easily recyclable. However, a simple pump bottle can be quite complicated to recycle: different plastics can be used to make up the bottle, including metal springs, and some components are even blends of plastic to ensure that the packaging is robust enough for formula shelf life. Glass may seem like the next best thing, since the material is less complicated to recycle. But, this material is not best suited for all types of skincare. Not to mention, glass actually doesn't have a great carbon footprint when it comes to transportation.

Non-recyclable products

The following is a general list of products that you typically won't be able to recycle due to size, incorporation of multiple materials, and incorporation of multiple components. We recommend confirming with brands on best recycling practices for their packaging.

- Lipstick tubes
- Sheet mask wrappers and sample pouches
- Most sunscreen tubes
- A lot of cleanser tubes
- Dropper tips and bottle pump heads

The good news is some work is being done to improve recyclability and manage plastic use. Here are three major initiatives going on in the cosmetic world:

PCR plastic: Post-consumer recycled plastic is one way we can help reduce the use of new virgin plastic. The tricky part is that PCR plastic is significantly more expensive to use. We hope that in the future we can work toward an economical way to incorporate PCR plastic and reduce virgin plastic use. At the current moment, it's difficult to figure out who eats this cost: the packaging supplier, the brand, or the skincare user.

Mono-material packaging: Mono-material packaging means all components of the skincare bottle are the same single type of plastic. This might sound counterintuitive to reducing plastic consumption, however, making a bottle mono-material greatly increases the chance that the bottle will be properly recycled and reused.

Major retail initiatives: A few retailers are doing their part in the recyclability challenge by creating initiatives to take in your beauty packaging waste whole and properly break down these complex components for you. It's a great way to take the complexity out of doing it yourself. There are still some limitations on what you can bring in, such as sheet-mask wrappers and aerosols, but check with your local store.

 You may occasionally see this icon, the Green Dot. It doesn't mean the product is recyclable; it means the company has made financial contributions to recycling and sustainability efforts.

What do these icons mean?

Listed here are the seven types of plastic you can recycle, their designated recycling number, and common products that use this type of plastic. Again, not an absolute definition—check for the appropriate recycling label. Be sure to wash your empty containers before recycling.

 1. Polyethylene terephthalate (PET): Most common type of plastic. Typically used for bottles and jars like mouthwash and toner bottles.

 2. High-density polyethylene terephthalate (HDPE): Shampoo, body wash, or lotion.

 3. Polyvinyl chloride (PVC) & vinyl: Shampoo bottles or blister packs.

 4. Low-density polyethylene (LDPE): Squeeze tubes.

 5. Polypropylene (PP): Bottle caps and cosmetic jars.

 6. Polystyrene (PS): Not commonly recycled; check with your local recycling center for best practices.

 7. Other plastics: Not commonly recycled; seek out your local waste management for guidance here.

"So, why can't we go biodegradable?": Fun fact! All things are biodegradable—it's simply a question of how long it takes for the actual object to biodegrade. For skincare, it's quite difficult to incorporate biodegradable packaging, since you also need this packaging to hold up on the manufacturing line and keep your formula safe and stable.

SHOPPING CHANNELS

You can find skincare anywhere, and that's not always a good thing! From your classic department store counters to upscale boutiques and even farmers' markets, each shopping channel comes with its own quirks and advantages. Here, we'll give you some chemists' guidelines on things to think about when you're hunting down your next skincare haul.

Specialty Retail

This category covers dedicated beauty retailers such as Sephora and Ulta.

Recommended for: These stores tend to be great for moisturizers, serums, and active treatments. For ingredients like vitamin C and acids, you're much more likely to find the right concentration in this channel, even though products will be a bit pricier than at the drugstore.

We're lukewarm on: Not too psyched about OTC products like sunscreen and acne treatments here. You might find a sunscreen with better texture in this avenue. But other than your favorite luxe sunscreen, there aren't a lot of reasons to splurge in these categories when there are viable drugstore alternatives of the same level of efficacy.

Chemists' Take on Price Point

What counts as "ridiculously priced" may be a bit subjective. From the chemists' perspective, there are a gazillion factors behind product pricing. You can find solid performers with the right percentages of proven active ingredients under $50 and sometimes well under $50. In our opinion, products with well-designed clinical studies to prove that their unique active combinations deliver claimed skin benefits can justify a $100 price tag.

However, products often charge a pretty penny for that weighty, luxurious packaging and not necessarily for the contents of the pretty bottle. These may be fun, pampering pieces, but they will not be your skincare workhorses. We also simply don't see any good reason for super-luxe products in the $300-and-up neighborhood. We just haven't seen any of those products perform with ten times the efficacy of a well-formulated $30 product!

Amazon

The internet in general is the Wild West of skincare. Amazon is a particularly diverse channel where you'll encounter *everything* under the sun. You can find anything from suspiciously cheap products to straight chemical raw materials to odd, exotic finds.

Recommended for: Japanese sunscreen— if you can't stand the greasiness of your sunscreen, we highly recommend scoping out Asian products. They can use filters not yet approved by the FDA, and they make some fantastic, light textures. We do, however, recommend a quick ingredient-list check. Alcohol, fragrance, and octinoxate are common ingredients in Asian formulas. While these are not inherently bad ingredients, some of you may have these on your personal no-no lists.

Face oils: We generally like elegant blends of face oils for more well-rounded benefits for your skin. But for those of you just getting into face oils, Amazon is a great place to find single-source oils you can try out to see which one your skin likes best.

Not recommended: Stay away from highly aggressive products. You can find absurd high-level actives on Amazon (70% glycolic acid! yikes!) or suspiciously cheap microneedle rollers. If it penetrates, or is positioned to remove warts, it's probably a terrible idea.

 Pro Tip: Look for sellers with transparent marketing claims and ingredient lists. Some may sell older products from past seasons, so be on the lookout for clearly labeled expiration dates! This is especially important for sunscreens!

When Is It Time to Toss?

 How do you know when to throw out your skincare? Check the label! All products should come with *this symbol* and should tell you how long you can use the product after opening:

Signs it's time to toss that skincare!

Of course, it can be hard to remember just when you started using a product. Look for these "uhh ohh" visual signs for when to toss out your product. If your skincare exhibits any of these visuals, RIP.

Drastic Color Change: Skincare formulas turning from white to a faint yellow are pretty normal, especially those with specialized actives. Formulas turning from a lovely white to dark brown is not!

Formula Separation: Chemists work hard to make sure your beautiful creams and lotions stay nicely uniform. If you see your product start to separate and you see clear liquid ooze out along with your white cream, that's a clear sign the formula is dying a sad death. We see this happen a lot with sunscreens.

Clear Formulas Are Not-So-Clear: You may find that your clear toners and sprays suddenly don't seem that clear. It's either gone cloudy or you may even find crystalized sediments.

Crystals: If you feel your serums, lotions, and creams suddenly have a curdled or exfoliating aspect to the texture and they're not labeled as scrubs, something has crystalized out and it's time to say goodbye.

Scent: Another telltale sign that your product is past its prime is any changes in scents. Fragrance may fade to some degree over time, but it should not be evolving into a whole different scent type. Fun fact! Some skincare actives can come with funky scents especially in the naturals realm. The degree of "funk" growing over time can be an indication that the active ingredient is losing its potency.

Skincare Tools and Devices

You can find many types of skincare gadgets and tools regardless of which shopping channel you turn to. The devices category seems to have exploded in recent years. From simple sponges to LED devices, you can find all sorts of tools to supplement your routine:

Cleansing tools: Perhaps one of the most common types of tools you can find on the market. These range from a simple konjac sponge to a high-tech electric brush. We're fans of devices in this category! They are a good way to amplify cleansing power without turning to harsher surfactants.

Massage tools: Massage tools like jade rollers and *gua sha* scrapers have skyrocketed in popularity in the past few years. Consider these as nice pampering tools, and take the fancy claims with a pinch of salt.

Light-therapy devices: There's a good amount of data out there on blue-light and infrared-light efficacy. As of the writing of this book, however, there isn't a home device we recommend that can deliver results as good as in-office, clinical devices.

Drugstore

This category covers CVS, Walgreens, and other similar stores.

Recommended for: Basic moisturizers and cleansers. You can also find great OTC products like sunscreen and acne treatments in this channel. This is an especially good shopping channel for all your nonprescription acne needs. Most of these places come with a full wall of salicylic acid cleansers, benzoyl peroxide treatments, and adapalene gels to choose from. OTC is highly regulated, so a 2% drugstore salicylic acid product wouldn't be less effective than a department store version by any means.

Not recommended: On the flip side, if you're looking for higher-strength, non-OTC active ingredients such as vitamin C or glycolic acid, the drugstore won't be the best place to find effective products. These ingredients are typically used on the much lower end, if not below, ideal concentration levels. At best, consider this a place for beginner-strength products for categories like retinol. Some detective work is required up front to find the good ones.

Asia-based E-Commerce Websites: Stylevana, Yesstyle

If you're an Asian skincare junkie or have been missing brands you can only get in the Far East, you're probably well aware of websites like Stylevana and Yesstyle.

Recommended for: This shopping channel offers the widest variety of Asia-based sunscreens, which means access to a lot of international filters and textures that you can't physically get in the US. Another advantage is that a lot of Asian brands formulate for very humid climates or layer-friendly routines. So if you're an oily-skin type that's looking for the lightest of hydration products, this is a good realm to explore.

Not recommended: It's pretty difficult to find treatments with higher concentrations of actives. A lot of Asian brands are a lot more conservative with their actives levels and they are also not the most transparent about their percentages. It's very difficult to know how much of their actives and extracts they're using in their skincare.

THE PROFESSIONALS

Our chemist perspective can help you make sense of the ingredients going into your skincare and how to organize all your skincare products. But this only provides one pillar of expertise, meaning there are a couple of other experts who can also really enhance your skincare journey. These experts can help your skincare go beyond at-home products and help heal your skin when it's having its worst day.

The Aesthetician

Consider an aesthetician as your face's therapist. Aestheticians work with skin day in and day out, and consistently seeing an aesthetician is like having a skin fairy-godmother to watch over your skin's progression. They can provide anything from simple facials to heavy-duty problem-solving procedures, such as deeper peels and microneedling.

Reasons to Visit Them:

• Very much-needed me time
• Seasonal check-in—stubborn blackhead extraction, anyone?
• High-strength peels and procedures

Things to Know Before Seeing an Aesthetician:

• Aestheticians may push products they believe in. One product is okay; trying to get you to buy a whole system? Not so much.
• Not every aesthetician is licensed to do all the procedures you may be interested in trying, so definitely do a little research if you have a specific procedure you've been wanting to try out.

The Dermatologist

Do you need guidance on serious acne? Do you suspect you may have an underlying skin condition? Or maybe you're just trying to figure out your pregnancy skincare routine. Welp, you've trekked into derm territory. Dermatologists will help diagnose and provide tailored prescription treatments for skin conditions such as eczema, psoriasis, acne, rosacea, and more. They are especially helpful in building a skincare routine while managing a skin condition. Even if you don't have a skin condition that calls for a dermatologist, you should be getting an annual checkup to keep tabs on your moles. The number of new cases of skin cancer being diagnosed is increasing, people!

Reasons to Visit Them:
- Struggling with any skin condition, including eczema, psoriasis, rosacea, and the like
- Acne breakouts
- Skin cancer checkup
- Invasive therapies such as lasers and injections

Things to Know Before Seeing a Dermatologist:
- Make sure the derm knows your full skincare routine and skin history.
- For first-time acne visits, bring in the products you're using so your derm has an idea of your current routine.

"Who knew there's a whole team of experts I can go to with my skin troubles!"

The Chemist

Well, of course we have to share about our chemist expertise. Chemists are the insiders on all things skincare products and skincare ingredients. Chemists do a lot more than just formulate what goes inside all these pretty bottles and jars. They actually do a lot of troubleshooting and quality control to make sure the formulas are safe to use for the duration of their shelf life. That means chemists have to have a lot of intimate knowledge on the ingredients they work with, and the regulations and restrictions around certain ingredients. They also have to make sure the formula and the packaging are compatible, and that the formula is preserved correctly to endure being in contact with human hands, face, and eyeballs.

Reasons to Consult a Chemist:
- Want to learn about a specific ingredient and if it's suited for your skin type
- Want to know what the difference is between a toner and a serum
- Want to know if you should be afraid of a preservative that just got some bad press
- Want to troubleshoot or optimize your skincare routine based on your skin goals.

Things to Know Before Talking to a Chemist:
- Don't ask them about any medical conditions or medically related issues. They are not doctors and shouldn't give any sort of diagnosis.
- Most are disgruntled and don't get enough love. If you luck out and are able to consult a chemist, ask nicely! And tip them...

GLOSSARY

active ingredients: Or just "actives," are what we chemists call key ingredients that deliver the claimed skincare benefits.

alpha hydroxy acids: Or simply AHAs, is a category of ingredients that function as chemical exfoliants. Key ingredients in this category are glycolic acid, lactic acid, and mandelic acid.

antioxidants: Substances that protect you from free radical damage.

barrier function: Your skin's all-important job of keeping irritants out and water in. Many products are formulated to support your barrier function.

beta hydroxy acid: BHA actually refers to salicylic acid. It is an effective acne ingredient.

broad spectrum: Indication that a sunscreen offers UVA protection.

ceramide: One of the key types of ingredients found in the lipid matrix of your stratum corneum. Also used in creams to improve hydration and bolster barrier function.

chemical exfoliant: An ingredient that chemically breaks down the bond between keratinocytes to slough off past-due skin cells.

chemical sunscreen: Also known as organic sunscreen, this includes a wide range of ingredients globally. Chemical sunscreens tend to have better texture but have controversial health and environmental implications.

clean beauty: A marketing term suggesting a product is more safe and "free from" allegedly unsafe ingredients. There is no standardized guideline of "clean," and the guidelines are determined by brands and retailers. This label does not actually imply that a product is more safe than others. Most ingredients flagged as "not clean" are often misunderstood or not actually used often in skincare.

collagen: One of the most important structural components of skin. Topical collagen acts as a hydrator.

dermis: The layer of skin below the epidermis, where collagen and elastin reside.

elastin: Another key protein responsible for that "bounce-back" in skin.

emollient: A type of ingredient found in moisturizers that works by smoothing skin. Plant oils are common emollients.

epidermis: The top layer of skin, where the stratum corneum and melanocytes reside.

free radicals: Reactive molecules that can cause collagen and DNA damage.

humectant: A type of ingredient found in moisturizers that works by grabbing hold of water to hydrate skin. Hyaluronic acid and glycerin are common humectants found in skincare products.

keratinocyte: The main skin cell in the epidermis. A keratinocyte located in the stratum corneum is also referred to as a corneocyte.

mechanical exfoliant: An ingredient that mechanically buffs the skin to physically remove past-due skin cells.

melanocyte: A cell in the inner layers of the epidermis in charge of producing melanin.

mineral sunscreen: Also known as physical sunscreen. There are just two ingredients in this category: zinc oxide and titanium dioxide. Recommended system for those with sensitive skin. Can leave undesirable white cast on skin.

occlusive: A type of ingredient found in moisturizers that works by protecting the skin from the elements. Petrolatum, shea butter, and beeswax are common occlusives.

OTC: Stands for "over the counter." This category of products has FDA-approved efficacy and stricter standards. Sunscreens, acne treatments, and skin protectants are the main categories of OTC products used in skincare.

polyhydroxy acids: Or simply PHAs, are newer chemical exfoliants on the market. They are considered gentler alternatives to AHAs, but there is less data on their efficacy.

retinoids: The umbrella term referring to the vitamin A ingredient family. Includes ingredients like tretinoin, retinol, retinaldehyde, and adapalene.

SPF value: A number that indicates the degree of UVB protection: 30 to 50 is the ideal target range that strikes the balance between sufficient protection and good skin texture.

stratum corneum: The outermost layer of the epidermis. This is your first line of defense against outside elements.

surfactant: A cleansing ingredient with a water-loving head and a fat-loving tail. This structure allows them to interact with the dirt and grime on your face as well as with water.

TEWL: Pronounced like *tool* and stands for "transepidermal water loss." This is a measurement of your skin health—the higher the TEWL, the worse your skin-barrier function.

The Big Four: The tried-and-true actives that have decades of research behind them for a variety of long-term skin benefits. These include chemical exfoliants, retinol, ascorbic acid, and niacinamide. Using a couple actives within the Big Four makes for a well-rounded anti-aging strategy.

tyrosinase: An enzyme in melanocytes that dictates how fast melanin is produced. Many whitening products work by slowing down tyrosinase.

BIBLIOGRAPHY

1. Grinnell, Frederick. "Fibroblast biology in three-dimensional collagen matrices." Trends in Cell Biology 13, no. 5 (2003): 264-269.

2. Schikowski, Tamara, and Anke Hüls. "Air pollution and skin aging." Current Environmental Health Reports (2020): 1-7.

3. Gfatter, R., P. Hackl, and F. Braun. "Effects of soap and detergents on skin surface pH, stratum corneum hydration, and fat content in infants." Dermatology 195, no. 3 (1997): 258-262.

4. Koski, Nina, E. Henes, C. Jauquet, Katy Wisuri, S. Rapaka, and L. Tadlock. "Evaluation of a sonic brush, cleanser, and clay mask on deep pore cleansing and appearance of facial pores through a new image analysis software methodology." Journal of the American Academy of Dermatology 70, no. 5 (2014): AB16.

5. Lampe, Marilyn A., A. L. Burlingame, JoAnne Whitney, Mary L. Williams, Barbara E. Brown, Esther Roitman, and Peter M. Elias. "Human stratum corneum lipids: Characterization and regional variations." Journal of Lipid Research 24, no. 2 (1983): 120-130.

6. Wilhelm, Klaus-P., Marianne Brandt, and Howard I. Maibach. "13 Transepidermal Water Loss and Barrier Function of Aging Human Skin." Bioengineering of the Skin: Water and the Stratum Corneum (2004): 143. Abingdon, UK: CRC Press.

7. Rawlings, A. V., David A. Canestrari, and Brian Dobkowski. "Moisturizer technology versus clinical performance." Dermatologic Therapy 17 (2004): 49-56.

8. Kligman, Albert M. "Petrolatum is not comedogenic in rabbits or humans: A critical reappraisal of the rabbit ear assay and the concept of acne cosmetica." Journal of the Society of Cosmetic Chemists 47.1 (1996): 41-48.

9. Pedersen, L. K., and G. B. E. Jemec. "Plasticising effect of water and glycerin on human skin in vivo." Journal of Dermatological Science 19, no. 1 (1999): 48-52.

10. Pavicic, Tatjana, Gerd G. Gauglitz, Peter Lersch, Khadija Schwach-Abdellaoui, Birgitte Malle, Hans Christian Korting, and Mike Farwick. "Efficacy of cream-based novel formulations of hyaluronic acid of different molecular weights in anti-wrinkle treatment." Journal of Drugs in Dermatology: JDD 10, no. 9 (2011): 990-1000.

11. Schroeder, P., C. Calles, T. Benesova, F. Macaluso, and J. Krutmann. "Photoprotection beyond ultraviolet radiation-effective sun protection has to include protection against infrared A radiation-induced skin damage." Skin Pharmacology and Physiology 23, no. 1 (2010): 15-17.

12. Wilson, Brummitte Dale, Summer Moon, and Frank Armstrong. "Comprehensive review of ultraviolet radiation and the current status on sunscreens." The Journal of Clinical and Aesthetic Dermatology 5, no. 9 (2012): 18.

13. Adler, Brandon L., and Vincent A. DeLeo. "Sunscreen safety: A review of recent studies on humans and the environment" Current Dermatology Reports (2020): 1-9.

14. Matta, Murali K., Robbert Zusterzeel, Nageswara R. Pilli, Vikram Patel, Donna A. Volpe, Jeffry Florian, Luke Oh et al. "Effect of sunscreen application under maximal use conditions on plasma concentration of sunscreen active ingredients: A randomized clinical trial." Journal of the American Medical Association 321, no. 21 (2019): 2082-2091.

15. Stamford, Nicholas PJ. "Stability, transdermal penetration, and cutaneous effects of ascorbic acid and its derivatives." Journal of Cosmetic Dermatology 11, no. 4 (2012): 310-317.

16. Bickers, David R., and Mohammad Athar. "Oxidative stress in the pathogenesis of skin disease." Journal of Investigative Dermatology 126, no. 12 (2006): 2565-2575.

17. Espinal-Perez, Liliana Elizabeth, Benjamin Moncada, and Juan Pablo Castanedo-Cazares. "A double-blind randomized trial of 5% ascorbic acid vs. 4% hydroquinone in melasma." International Journal of Dermatology 43, no. 8 (2004): 604-607.

18. Barnes, M. J. "Function of ascorbic acid in collagen metabolism." Annals of the New York Academy of Sciences 258, no. 1 (1975): 264-277.

19. Ali, Basma M., Amal A. El-Ashmawy, Gamal M. El-Maghraby, and Rania A. Khattab. "Assessment of clinical efficacy of different concentrations of topical ascorbic acid formulations in the treatment of melasma." Journal of the Egyptian Women's Dermatologic Society 11, no. 1 (2014): 36-44.

20. Pinnell, Sheldon R. "Cutaneous photodamage, oxidative stress, and topical antioxidant protection." Journal of the American Academy of Dermatology 48, no. 1 (2003): 1-22.

21. Tagami, Hachiro. "Functional characteristics of the stratum corneum in photoaged skin in comparison with those found in intrinsic aging." Archives of Dermatological Research 300, no. 1 (2008): 1-6.

22. Kornhauser, Andrija, Sergio G. Coelho, and Vincent J. Hearing. "Applications of hydroxy acids: Classification, mechanisms, and photoactivity." Clinical, Cosmetic and Investigational Dermatology: CCID 3 (2010): 135.

23. Bernstein, Eric F., Douglas B. Brown, Mark D. Schwartz, Kays Kaidbey, and Sergey M. Ksenzenko. "The polyhydroxy acid gluconolactone protects against ultraviolet radiation in an in vitro model of cutaneous photoaging." Dermatologic surgery 30, no. 2 (2004): 189-196.

24. DiNardo Joseph C., Gary L. Grove, and Lawrence S. Moy. "Clinical and histological effects of glycolic acid at different concentrations and pH levels." Dermatologic Surgery 22, no. 5 (1996): 421-424.

25. Stiller, Matthew J., John Bartolone, Robert Stern, Shondra Smith, Nikiforos Kollias, Robert Gillies, and Lynn A. Drake. "Topical 8% glycolic acid and 8% L-lactic acid creams for the treatment of photodamaged skin: A double-blind vehicle-controlled clinical trial." Archives of Dermatology 132, no. 6 (1996): 631-636.

26. Mekas, Maria, Jennifer Chwalek, Jennifer MacGregor, and Anne Chapas. "An evaluation of efficacy and tolerability of novel enzyme exfoliation versus glycolic acid in photodamage treatment" Journal of Drugs in Dermatology: JDD 14, no. 11 (2015): 1306-1319.

27. Draelos, Zoe Diana, Keith D. Ertel, and Cynthia A. Berge. "Facilitating facial retinization through barrier improvement." Cutis, no. 4 (2006): 275.

28. Draelos, Zoe Diana, Akira Matsubara, and Kenneth Smiles. "The effect of 2% niacinamide on facial sebum production." Journal of Cosmetic and Laser Therapy 8, no. 2 (2006): 96-101.

29. Hakozaki, T., L. Minwalla, J. Zhuang, M. Chhoa, A. Matsubara, K. Miyamoto, A. Greatens, G. G. Hillebrand, D. L. Bissett, and

R. E. Boissy. "The effect of niacinamide on reducing cutaneous pigmentation and suppression of melanosome transfer." British Journal of Dermatology 147, no. 1 (2002): 20-31.

30. Camargo Jr. Flávio B., Lorena R. Gaspar, and Patricia MBG Maia Campos. "Skin moisturizing effects of panthenol-based formulations." Journal of cosmetic science 62, no. 4 (2011): 361.

31. Pinnock, Carole B., and Christopher P. Alderman. "The potential for terato-genicity of vitamin A and its congeners." Medical journal of Australia 157, no. 11 (1992): 804-809.

32. Kong, Rong, Yilei Cui, Gary J. Fisher, Xiao-juan Wang, Yinbei Chen, Louise M. Schnei-der, and Gopa Majmudar. "A comparative study of the effects of retinol and retinoic acid on histological, molecular, and clinical properties of human skin." Journal of cosmetic dermatology 15, no. 1 (2016): 49-57.

33. Mukherjee, Siddharth, Abhijit Date, Vandana Patravale, Hans Christian Korting, Alexander Roeder, and Günther Weindl. "Retinoids in the treatment of skin aging: an overview of clinical efficacy and safety." Clinical Interven-tions on Aging 1, no. 4 (2006): 327.

34. Fisher, Gary J., and John J. Voorhees. "Molecular mechanisms of retinoid actions in skin." The FASEB Journal 10, no. 9 (1996): 1002-1013.

35. Dhaliwal, S., I. Rybak, S. R. Ellis, M. Notay, M. Trivedi, W. Burney, A. R. Vaughn et al. "Prospective, randomized, double blind assessment of topical bakuchiol and retinol for facial photoageing." British Journal of Dermatology 180, no. 2 (2019): 289-296.

36. Haftek, Marek, Sophie Mac Mary, Marie Aude Le Bitoux, Pierre Creidi, Sophie Seité, André Rougier, and Philippe Humbert. "Clinical, biometric and structural evaluation of the long-term effects of a topical treatment with ascorbic acid and madecassoside in photoaged human skin." Experimental Dermatology 17, no. 11 (2008): 946-952.

37. Lee, J., H. Jun, E. Jung, J. Ha, and D. Park. "Whitening effect of bisabolol in Asian women subjects." International Journal of Cosmetic Science 32, no. 4 (2010): 299-303.

38. Beitner, Harry. "Randomized, placebo controlled, double blind study on the clinical efficacy of a cream containing 5% lipoic acid related to photoageing of facial skin." British Journal of Dermatolo-gy 149, no. 4 (2003): 841-849.

39. Knott, Anja, Volker Achterberg, Christoph Smuda, Heiko Mielke, Gabi Sperling, Katja Dunckelmann, Alexandra Vogelsang et al. "Topical treatment with coenzyme Q 10-containing formulas improves skin's Q 10 level and provides antioxidative effects." Biofactors 41, no. 6 (2015): 383-390.

40. Tzung, Tien Yi, Kuan Hsing Wu, and Mei Lun Huang. "Blue light phototherapy in the treatment of acne." Photodermatology, Photoimmunology & Photomedicine 20, no. 5 (2004): 266-269.

41. Çetiner, Salih, Turna Ilknur, and Ebnem Özkan. "Phototoxic effects of topical azelaic acid, benzoyl peroxide and adapalene were not detected when ap-plied immediately before UVB to normal skin." European Journal of Dermatology 14, no. 4 (2004): 235-237.

42. Zolghadri, Samaneh, Asieh Bahrami, Mahmud Tareq Hassan Khan, J. Munoz-Munoz, Francisco Garcia-Molina, F. Garcia-Canovas, and Ali Akbar Saboury. "A comprehensive review on tyrosinase inhibitors." Journal of Enzyme Inhibition and Medicinal Chemistry 34, no. 1 (2019): 279-309.

43. Ebrahimi, Bahareh, and Farahnaz Fatemi Naeini. "Topical tranexamic acid as a promising treatment for melasma." Journal of Research in Medical Sciences 19, no. 8 (2014): 753.

44. Uitto, Jouni. "The role of elastin and collagen in cutaneous aging: Intrinsic aging versus photoexposure." Journal of Drugs in Dermatology: JDD 7, no. 2 Suppl (2008): s12.

45. Ellis, Millikan, Smith, Chalker, Swinyer, Katz, Berger, et al. "Comparison of adapalene 0.1% solution and tretinoin 0.025% gel in the tropical treatment of acne vulgaris" British Journal of Dermatology 139, s52 (1998): 41–47.

46. Thiboutot, Diane M., Jonathan Weiss, Alicia Bucko, Lawrence Eichenfield, Terry Jones, Scott Clark, Yin Liu, Michael Graeber, and Sewon Kang. "Adapalene–benzoyl per-oxide, a fixed-dose combination for the treatment of acne vulgaris: Results of a

multicenter, randomized double-blind, controlled study" Journal of the Amer-ican Academy of Dermatology 57, no.5 (1998): 791–99.

47. Williams, Hywel C. Robert P. Dellavalle, and Sarah Garner. 2012. "Acne vulgaris." Lancet 379, n. 9813 (2012) 361–72.

44. Huang, Y.-L., Chang, S.-L., Ma, L., Lee, M.-C., & Hu, S. (2013). Clinical analysis and classi-fication of dark eye circle. International Journal of Dermatology, 53(2), 164–170. doi:10.1111/j.1365-4632.2012.05701.x .

45. Seidel R, Moy RL. Reduced appearance of under-eye bags with twice-daily appli-cation of epidermal growth factor (EGF) serum: a pilot study. Journal of Drugs in Dermatology : JDD. 2015 Apr;14(4):405-410. PMID: 25844616.7.

46. Dayal, S., Sahu, P., Jain, V.K. and Khetri, S. (2016), Clinical efficacy and safety of 20% glycolic peel, 15% lactic peel, and topical 20% vitamin C in constitutional type of periorbital melanosis: a compar-ative study. J Cosmet Dermatol, 15: 367-373. https://doi.org/10.1111/jocd.12255

47. Ahmadraji F, Shatalebi MA. Evaluation of the clinical efficacy and safety of an eye counter pad containing caffeine and vitamin K in emulsified Emu oil base. Adv Biomed Res. 2015 Jan 6;4:10. doi: 10.4103/2277-9175.148292. PMID: 25625116; PMCID: PMC4300604.

INDEX

ABOUT THE AUTHORS

GLORIA studied chemical engineering at Cornell University with a focus on sustainable energy. She ended up in the beauty industry serendipitously after landing an internship in L'Oreal's lipstick lab, which then turned into a full-time gig as a skincare-formulation chemist.

VICTORIA studied chemical engineering at The University of California at San Diego and went on to earn a master's degree in nanoengineering, researching nanotherapeutics for acne. She had no intention of entering pharma and was lucky enough to land a job in L'Oreal's skincare lab and ended up being Gloria's cubicle neighbor.

Nowadays, these two spend their time juggling in tandem the 17 hats needed to run their startup, Chemist Confessions. In Gloria's spare time, she hangs out with her two cats, and Victoria tries very hard to keep her plants alive.

ACKNOWLEDGMENTS

We expected a roller-coaster ride with Chemist Confessions but would never have guessed this adventure would lead us here. A real book. We would like to thank everyone who helped make this a reality. Thanks for catching typos, giving us much-needed pep talks, and checking in on us during this mountain of a project. **THANK YOU.**

Support Circle: Moms, Dads, Lisa, Kath, Ryan, Chris

Integral Proofreaders: Adam, Erica, Irene, Rosie, Celia

Pet Therapists: Kuma, Hana, Rae, Kalin, Boe, Maisy, Roxy

weldon**owen**

an imprint of Insight Editions
P.O. Box 3088
San Rafael, CA 94912
www.weldonowen.com

CEO Raoul Goff
VP Publisher Roger Shaw
Editorial Director Katie Killebrew
Senior Editor Karyn Gerhard
Editorial Assistant Jon Ellis
VP Creative Chrissy Kwasnik
Art Director Allister Fein
VP Manufacturing Alix Nicholaeff
Sr Production Manager Joshua Smith
Sr Production Manager, Subsidiary Rights
Lina s Palma-Temena

Weldon Owen would also like to thank
Margaret Parrish for proofreading and
Kevin Broccoli of BIM for the index.

Photography by Lorena Masso

All illustrations by Victoria Fu with the
following exceptions:

Patterns used throughout and images on pages
16-21, 25, 34, 48-51, 72, 75-78, 84, 95, 116, 132, 134,
140, 148, 156, 206, and 223 by Adam Raiti

Portraits of the authors by Lorraine Rath

Icons courtesy of Shutterstock

Revised and expanded edition
© 2024 Weldon Owen
First edition published 2021

ISBN: 979-8-88674-241-1

Manufactured in China by Insight Editions
10 9 8 7 6 5 4 3 2 1

ROOTS of PEACE   REPLANTED PAPER

Insight Editions, in association with Roots of Peace,
will plant two trees for each tree used in the
manufacturing of this book. Roots of Peace is an
internationally renowned humanitarian organization
dedicated to eradicating land mines worldwide and
converting war-torn lands into productive farms
and wildlife habitats. Roots of Peace will plant two
million fruit and nut trees in Afghanistan and provide
farmers there with the skills and support necessary
for sustainable land use.